Python for Microcontrollers

Getting Started with MicroPython

Donald Norris

New York   Chicago   San Francisco
Athens   London   Madrid
Mexico City   Milan   New Delhi
Singapore   Sydney   Toronto

Library of Congress Control Number: 2016954567

Python for Microcontrollers: Getting Started with MicroPython

1 2 3 4 5 6 7 8 9   LCR   21 20 19 18 17 16

ISBN   978-1-25-964453-5
MHID   1-25-964453-7

This book is printed on acid-free paper.

Sponsoring Editor	**Project Manager**	**Indexer**
Michael McCabe	Anju Joshi, Cenveo® Publisher Services	Robert Swanson
Editorial Supervisor		**Art Director, Cover**
Stephen M. Smith	**Copy Editor**	Jeff Weeks
Production Supervisor	Lisa McCoy	**Illustration**
Lynn M. Messina	**Proofreader**	Cenveo Publisher Services
Acquisitions Coordinator	Barnali Ojha, Cenveo Publisher Services	**Composition**
Lauren Rogers		Cenveo Publisher Services

This book is dedicated to my three wonderful grandchildren,
Hudson and Holton Norris and Evangeline Kachavos.
They are an absolute joy to be with, although I am sure
that sentiment is not always shared by their parents.

About the Author

Donald Norris has a degree in electrical engineering and an MBA specializing in production management. He is currently teaching undergrad and grad courses in the IT subject area at Southern New Hampshire University. He has also created and taught several robotics courses there. He has over 36 years of teaching experience as an adjunct professor at a variety of colleges and universities.

Mr. Norris retired from civilian government service with the U.S. Navy, where he specialized in acoustics related to nuclear submarines and associated advanced digital signal processing. Since then, he has spent more than 22 years as a professional software developer using C, C#, C++, Python, Node.js, and Java, as well as 5 years as a certified IT security consultant.

Mr. Norris started a consultancy, Norris Embedded Software Solutions (dba NESS LLC), that specializes in developing application solutions using microprocessors and microcontrollers. He likes to think of himself as a perpetual hobbyist and geek, and is always trying out new approaches and out-of-the-box experiments. He is a licensed private pilot, photography buff, amateur radio operator, and avid runner.

Besides the present book, Mr. Norris is the author of *Raspberry Pi® Projects for the Evil Genius™*, *Programming the Intel® Edison: Getting Started with Processing and Python*, and three other TAB books.

CONTENTS

PREFACE

This is the first book written on the MicroPython language. Although several good tutorials are available on the Web, especially at www.micropython.org, I have found that a book format is often an attractive alternative for readers who do not wish to invest the time and energy needed to wade through online tutorials. In addition, and without disparaging online authors, a well-written and -organized book often provides readers a good path to understanding a new and innovative technology such as MicroPython.

I have also presented a few projects in this book, which should interest most readers. There is one on a robotic car, another on building a global positioning system (GPS) sentence decoder, two on how to use Pyboard skins, and one on a let ball detector for those readers who play tennis. The theme of all of the projects is to learn how to interface a MicroPython microcontroller with external devices and sensors.

I cover three separate modules that run MicroPython in this book. The first is the Pyboard, which was designed by Dr. Damien George, who is also the creator of the MicroPython language. All of the introductory scripts are run on the Pyboard, as well as all of the book's projects. The other two boards discussed are the ESP8266 and the WiPy. These boards run MicroPython, but they also have a built-in wireless radio system, which is lacking in the Pyboard. In fact, the WiPy has four separate wireless radio systems, which makes it quite a chameleon if you need a module that can connect to a variety of wireless protocols. Learn all about it in Chapter 10.

The reader should expect a great increase in knowledge of and experience with MicroPython if a commitment is made to read all of the introductory material as well as to complete most of the projects. I personally always learn a great deal while designing and finishing them. Often, things work out just fine, whereas at other times they are fraught with problems. However, that's what I consider the joy of experimenting. As the renowned Professor Einstein once stated, "Anyone who has never made a mistake has never tried anything new."

I would caution any experienced Linux developers to at least review the beginning chapters because MicroPython, although modeled after Python 3, does have a few "gotchas" that one should be aware of before wading too deeply into the MicroPython pool. I have tried to provide many useful hints and techniques throughout the book to assist readers in this MicroPython adventure.

So, without further ado, let the adventure begin.

Donald Norris

ACKNOWLEDGMENTS

I first thank my lovely wife, Karen, for putting up with all my experiments and enduring all the "discussions" about the book projects.

Thanks go to Michael McCabe for his fine support as my editor and also to Amy Stonebraker for her support as editorial assistant.

In addition, thanks go out to Anju Joshi for her fine work as the project manager.

Finally, I would like to thank Dr. Damien George at www.micropython.org for his excellent efforts in creating the MicroPython language and getting the Pyboard out to the marketplace.

1

Introduction

This book is about the MicroPython language; it includes eight projects using the Pyboard, the ESP8266 or the WiPy board. Projects using one of these boards will be discussed in great detail in later chapters; however, I will be using the Pyboard to introduce the MicroPython language because it comes installed in it and ready to use.

Introduction to MicroPython

MicroPython is a variant of the Python 3 language designed specifically to run on memory-constrained microcontrollers. It was the idea of Dr. Damien P. George of Cambridge University, who first created an initial version in 2013 along with a Kickstarter campaign to fund the production of the Pyboard, which is a specially designed hardware development board hosting MicroPython as an operating system (OS). Dr. George subsequently presented MicroPython, version 1.0, at a 2014 Python Conference (PyCon). Dr. George also started the website micropython.org to support both the language and the Pyboard. All the latest information regarding the MicroPython language and the Pyboard may be found on this website. There are also several very active user forums hosted on this site, which I found to be very beneficial in answering questions I had on MicroPython and the Pyboard.

Design Philosophy

At the outset, I would like to disclose my design philosophy regarding the proper use of MicroPython and the Pyboard. You should note that this disclosure only represents my philosophy and not necessarily Damien George's approach, although I would suspect that they are not too different.

MicroPython as implemented on the Pyboard is designed to quickly and efficiently create control programs for embedded projects. An embedded project can be simply defined as any project that requires the use of a microcontroller to meet project requirements, with or without human intervention. Typical embedded projects often use sensors and sometimes electromechanical actuators, which are interfaced to the microcontroller. They may have human interface devices (HIDs) attached to facilitate human interaction, but most often they are autonomous, acting either in a stand-alone capacity or linked to the cloud for remote access.

Embedded projects usually do not require or need general-purpose applications such as a terminal program or a database program. MicroPython does not easily support adding such high-level applications. You are likely using an inappropriate board if you find that you require these types of applications for your project. A Raspberry Pi or BeagleBone Black would likely be a much better fit for your project, as both support powerful Linux distributions that are easily adapted and configured to meet high-level requirements.

In summary, MicroPython and the Pyboard are best suited for rapid development of embedded applications using a high-level language. Such applications are very efficient when operating and use minimal memory compared with more traditional development approaches.

Exploring MicroPython

Before I start delving into the intricacies and fine details of both Python and MicroPython, I thought I would show both logos. Figure 1-1 shows the logo of the official Python language, which can only be formally associated with Python language implementations recognized and supported by the Python organization at python.org. It is also known informally as the two snakes logo.

Figure 1-1 *Official Python logo.*

Figure 1-2   *MicroPython logo.*

In contrast, the MicroPython logo shown in Figure 1-2 seems to be a bit more informal while at the same time reflecting the emphasis on microcontroller usage as reflected by the physical chip in the logo.

The interesting aspect is that the Python name is not based on the python snake species, but rather springs from Python creator Guido van Rossum's interest in the U.K. comedic group Monty Python.

In the next section I will present object-oriented (OO) concepts along with a discussion on the Python language describing its main features and its original intent. This will be necessary to understand why MicroPython was created and why it is truly an important addition to the Python community. Feel free to skip the next several sections if you are already quite comfortable with using Python and object-oriented technology and just really want to start using MicroPython.

Object-Oriented Programming and Some Python Basics

Let me first state that this section is not intended to be a tutorial on either OO or Python. There are many fine tutorials freely available on the Web covering both topics, which can provide a much better background than I could possibly provide. My intention is to provide a sufficient background in both object-orientation and key Python concepts to enable you to gain a good understanding of both what makes Python function so well and why you should use it for your microcontroller projects. You should read this material keeping in mind that if some concept is fuzzy or unclear it might behoove you to research it further before continuing with this book. Thankfully, no quizzes or exercises are associated with this section other than one incredibly easy question, which I will discuss shortly.

Object-Oriented Concepts

To begin our exploration of object orientation, let us pretend we have been transported to a virtual environment where objects are the primary life form. We will call this environment Object Land.

Figure 1-3 shows a very abstract view of Object Land where two processes are shown (small and big), each containing multiple objects that are in constant communication with one another to accomplish the overall process goals. The processes themselves are communicating with each other as needed to accomplish whatever needs to be done.

A key question arises: What is an object?

The textbook answer typically given is that an object is an instance of a class. Of course, this only further confuses the newcomer to Object Land where he (used only in the generic sense) doesn't know the definition of a class. OK, so what is the definition of a class and, more important, why should I care?

First, a quick quiz. What does Figure 1-4 represent?

- Bus

- Train

- Race car

- Plane

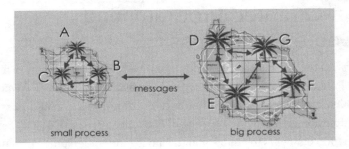

Figure 1-3   *Object Land.*

Figure 1-4 *Unknown object.*

The answer really lies with your life experience. Most people will know it is a race car by its shape and the fact that the driver is wearing a helmet. Others may recognize it by the process of elimination by realizing it is not a bus, train, or plane. We engage in this process continuously—that is, we use models or abstractions to represent real-world things or objects.

Similar activities are present in software design where we use abstractions to represent real-world things. This approach is much more relevant in developing software compared with a much stricter procedural approach. Consider a situation where you at the train station exit having just arrived in New York City. Now you want to go to Radio City Music Hall and take in a show, so you hail a cab. Once in the taxi, do you tell the driver, "Go to the end of the street; take a right; go through two sets of lights; take a left . . ." or do you simply say "Please take me to Radio City Music Hall"? The first approach is procedural, whereas the latter is object oriented. In taking the OO approach, you are relying on that person object (the taxi driver) to be responsible for accepting a message "Please take me to Radio City Music Hall" and know how to accomplish the task. One very nice feature of this approach is that the object may have to change its implementation depending on traffic, street closures, etc., but you as the message sender will not be aware of this change. You have to be aware of all the traffic conditions in New York City if you choose the procedural approach. Not a very appealing situation!

Having established the fact that objects will in fact be useful to accomplish your goals in controlling sensors, it is time to examine some fundamental principles underlying all object-oriented programming (OOP) paradigms. Table 1-1 lists the four bedrock principles that apply to all OOP languages.

Principle	Description
Abstraction	Modeling to represent real-world things
Polymorphism	Different object behavior generated by the same message
Inheritance	Shared common attributes and behaviors among objects
Encapsulation	Closing off the inner workings of objects from public view

Table 1-1 *Four Bedrock Object-Oriented Principles*

A handy mnemonic to remember these principles is APIE, taken from the beginning letter for each principle.

I decided to use a generic robot as a model to demonstrate how to apply the OO approach. Determining basic robot characteristics and behaviors is normally the first step in creating a class. The class is the data structure used to record these characteristics and behaviors. In formal OO terms, characteristics are known as attributes, and behaviors are methods. Objects are created from classes. As mentioned earlier, an object is simply an instance of a class. How this is done depends on the specific language being used. Many OO languages such as Java, C#, and C++ use the new operator to create a class instance. In Python, all you need to do is use the assignment operator or = symbol. This process is known as instantiation.

It is often useful to refer to basic definitions in developing class attributes and methods. Many robot definitions are available, but I have created one that fits my understanding and belief of what a robot is, and it is one that I believe you will find useful when trying to create a robot class definition: *a robot is a system with sensors, actuators, and a feedback control mechanism.* Note that I avoided the use of the words *knowledge* and *intelligence*. It is readily apparent that robot can appear intelligent; however, I want to side-step the whole artificial intelligence discussion for now. In addition, the word *actuators* implies the action or motion normally associated with robots. Some robots are quite mobile, whereas others are fixed in place, such as a robotic work cell. Additionally, the reference to a feedback control mechanism implies the use of (a) sensor(s) and the ability to react to the data generated by the sensor(s).

The key is to try to encapsulate all the essential attributes and behaviors that are useful in describing a real-world object in a logical data structure such as a class. I also wish to emphasize that there is really no one correct answer to creating a class. It turns out some descriptions are better than others, and you will find that as you proceed with your design you will often go back and revise your initial class definition. Experience in repeated OO design efforts will improve your initial efforts, and incorporating design patterns (DPs), which I discuss later, will also help with the design. I have repeatedly told my beginning OO students that creating classes are probably the single hardest task to tackle in the whole OO approach.

Modeling a Robot

In this section I will provide a simple, yet realistic, example of how to create a model of a generic robot, which provides a good template to model other, more specific robot types. Figure 1-5 shows a simple robot inheritance class diagram with a parent class and four child classes. The parent class contains the general attributes and methods common to all the child robot classes. Unified Modeling Language (UML), version 2, standards were followed in constructing Figure 1-5. UML is the software development industry's standard way of displaying graphical models. Knowing how to create useful UML diagrams promotes efficiency and effectiveness in communicating your design ideas to others in the development process.

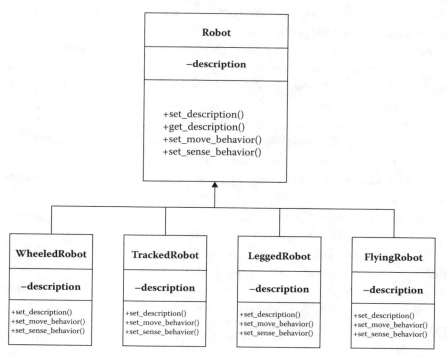

Figure 1-5 *Robot UML class diagram.*

The four child classes—namely Wheeled, Tracked, Legged, and Flying—get their names from their primary mode of movement. These four robot subclasses can be instantiated as they have specific implementations for the methods declared, but not implemented, in the parent class. Inheritance is very useful in promoting software reuse, but it does have its drawbacks. It must be used in a very careful approach to avoid situations where too many unique objects could be created by combining different parent case attributes and/or methods. Interfaces will be discussed later as a more elegant way of creating objects with limited use of inheritance.

The concept of scope is also important in OOP. Scope is the way OOP enforces encapsulation. Objects "know" things about themselves, that is, their attributes, and they also know how to do things, which includes their methods or behaviors. You don't want outside entities changing these properties without granting the entities permission. Scope enforces this constraint by setting attributes and methods as private, public, or protected. Private scope means exactly what its name states: attributes and methods are only available within the encompassing class and consequently all objects instantiated from that class. Outside entities cannot change, modify, or delete private attributes or methods. A minus sign (−) in front of a UML entry indicates private scope.

Attributes and methods marked as public are available to outside entities. However, attributes themselves are rarely made public because that typically destroys class encapsulation. Public methods, on the other hand, are the way classes allow messages to be both received and sent by class objects. This approach is termed the public interface, and for the vast majority of situations, is the way classes are created. In some advanced OOP areas there are inner classes that do not require a public interface to achieve the desired functionality. Inner classes will not be required for sensor programming. A plus sign (+) in front of a UML entry indicates public scope.

The final type of scope is protected. This is almost identical to private, except child objects are permitted access to parent attributes and methods declared protected, but no other outside entity is granted permission. Protected scope helps with inheritance class structure implementations, and child objects are always treated as if they are the same type as the parent. From our robot class diagram we can make the statement "A Wheeled or Tracked object is a Robot object." Inheritance is always the "is a" relationship.

Another key concept to consider is composition, which is a situation where a class contains attributes that are objects instantiated from other classes. Composition allows us to build complex objects, just as real-world things are made up of different components. Consider a car that contains an engine. You can easily imagine a Car class; however, it would be a big mistake to create a child class named Engine. It would fail the commonsense inheritance test of stating "an Engine is a Car," which is required for true inheritance to exist. However, you can state with confidence, "a Car has an Engine." Composition is the "has a" relationship. Composition and the closely related aggregation relationship concept are extremely helpful in creating useful and descriptive classes and are needed to successfully implement the interface concept mentioned earlier.

An interface is a specialized class that only contains methods and no attributes. Classes using interfaces can supplement the methods that are declared either within the class or within a parent class if inheritance is used. Interfaces also support inheritance to allow for specialized method implementations that fit specific subclass requirements. An example is really needed to clarify how interfaces are best used.

Some Python Basics

I felt it important to now cover some basic Python concepts before presenting some example code implementing the general OO principles I presented earlier. I will first start by stating that Python is a true OO language, strictly adhering to the APIE bedrock principles shown in Table 1-1. Sometimes, programmers become confused and misled by Python's interactive nature; but make no mistake—it is truly an OO language, capable of implementing most, if not all, of the sophisticated implementations that developers can dream up.

I will be using Linux running on a Raspberry Pi 2, Model B, for the sake of simplicity and recognizing that it is somewhat similar to a Pyboard. This particular Python version is 2.7.3, which obviously differs from the version 3.4 on which MicroPython is based. The differences for my discussion purposes are minimal, and I will point out where they exist. Figure 1-6 shows an interactive session in which I have input several different operations and assignments to demonstrate some Python actions.

At first you should note that I started the Python interpreter by simply entering `python` at the RasPi's command-line prompt. This brings up the Python

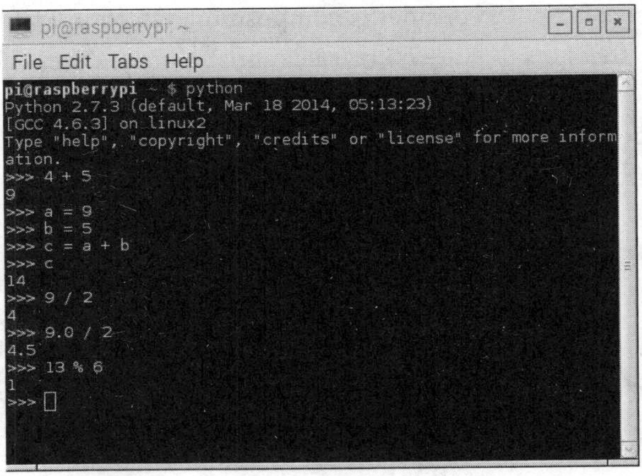

Figure 1-6 *Python interactive session.*

interactive prompt >>> along with some introductory words. I next entered a simple addition of two integer numbers, which was immediately executed after I pressed the ENTER (RETURN) key. Python will always try to execute any properly entered expression and return the results, if any, to the user.

The next entry shows how to allocate and initialize a variable—in this case, a variable named *a*, which is set to a value of 6. No return value is shown, nor is there any error message, which means that Python both accepted the variable assignment and its initialization. This simple step actually reveals something quite remarkable about Python, in that it is a strongly typed language and it invokes dynamic allocation. This means that Python dynamically created a variable named *a* that holds or stores integer (whole) numbers. I could have later on reassigned a value of 6.0 to *a*, which would have caused Python to change the *a* variable type from integer to floating point or real, that is, a number containing a decimal point. This action is known as dynamic typing and makes it easy for users to create programs or scripts without much forethought, although there are some inherent problems lurking if you are not careful.

The next few steps in the figure show how two more variables were created: one initialized to a hard value and the other the sum of the two previous variables. I then entered the last variable, *c*, and pressed ENTER. Its computed value was immediately displayed.

The next calculation demonstrates how Python's strong typing may confuse you. I entered the calculation 9 / 2 and was shown 4, not 4.5 as one might reasonably expect. This is because I asked Python to divide one integer 9 by a second integer 4, which it obediently did. The result is the integer 4, which is the whole number of times 2 may be divided into 9. There is no concept of fractional remainders in integer mathematics. To "fix" this problem, all I needed to do was change the integer 9 to a real number, 9.0, and repeat the calculation, and voila: 4.5 was the correct response.

The last calculation of 13 % 6 demonstrates integer remaindering where the % operator returns the remainder of the expression—in this case, 1. This % operator is also known as the modulo operator, and I find it quite useful for specific math calculations.

It would be impractical and very inefficient to only have the interactive mode available to Python users. Fortunately, Python allows for prewritten programs or scripts to be executed. These function precisely the same way as if a user entered them directly via the interactive interpreter. However, scripts offer more in that many predefined modules such as classes can be quickly referenced and instantiated, allowing for a fully featured OO development environment. From this point on, I will only be using Python programs to demonstrate the language and all its OO capabilities. Note that you should use any text editor you feel comfortable using as the means to enter the code shown in this book. I will be using the nano editor, which is readily available for installation on the RasPi by entering the following command:

```
sudo apt-get install nano
```

I also believe that nano already comes installed in the latest Raspian Linux distributions. You may also copy and paste the code from this book's companion website, www.mhprofessional.com/micropython.

The Robot Class

The following Python code defines my first attempt at modeling a generic robot using the guidelines I discussed earlier. There are only two key behaviors, move and sense, that are common to all the robot subclasses as defined in the UML model. Child classes that use these behaviors, or methods as I shall call them from now on, also provide specific method implementations suitable for their respective robot models.

Note: This code is *not* strictly compliant with the way Python purists recommend that OO code be implemented. However, I feel that it adequately conveys the OO principles I wish to discuss while keeping the material relatively straightforward and, hopefully, understandable.

```python
class Robot(object):

    def __init__(self):
      pass

    def set_description(self, desc):
            self.description = desc

    def get_description(self):
            pass

    def set_move_behavior(self, mb):
            self.move_behavior = mb

    def set_sense_behavior(self, sb):
            self.sense_behavior = sb

    def perform_move(self):
            self.move_behavior.move()

    def perform_sense(self):
      self.sense_behavior.sense()
```

I will next discuss two of the four child classes shown in Figure 1-5.

Child Classes

This section will demonstrate how to create WheeledRobot and TrackedRobot child classes. Print line statements will be used, as that is the simplest and most effective way of showing object behavior without using actual hardware.

I will also use two of the methods detailed in the Robot class diagram, as this will be sufficient to illustrate object interaction and behavior. These methods are:

* perform_move()

* perform_sense()

The WheeledRobot class definition follows.

```
from robot import Robot
from move_with_wheels import MoveWithWheels
from sense_with_ping import SenseWithPing
class WheeledRobot(Robot):
    def __init__(self):
        move_instance = MoveWithWheels()
        self.set_move_behavior(move_instance)
        sense_instance = SenseWithPing()
        self.set_sense_behavior(sense_instance)
        desc = "I am a robot with wheels"
        self.set_description(desc)
```

In Python, an inheritance relationship is created simply by placing the parent class name in parentheses after the child class name. In this situation, the statement:

```
class WheeledRobot(Robot):
```

creates the inheritance relationship that the Robot class is a parent of the child class WheeledRobot.

In many OO languages, there is a special method known as a constructor, which initializes an object's attributes to known and consistent states prior to use. Python does not use a constructor, per se, but instead uses the __init__ method, which is very similar in functionality to the constructor. In this class definition, the __init__ method associates the appropriate movement and sense functions to the object, which will be moving with wheels and sensing distance using a Ping ultrasonic sensor.

The TrackedRobot class is set up in a similar fashion to the WheeledRobot class and is shown next.

```
from robot import Robot
from move_with_tracks import MoveWithTracks
from sense_with_lidar import SenseWithLidar
class TrackedRobot(Robot):
    def __init__(self):
        move_instance = MoveWithTracks()
        self.set_move_behavior(move_instance)
        sense_instance = SenseWithLidar()
    self.set_sense_behavior(sense_instance)
        desc = "I am a robot with tracks"
        self.set_description(desc)
```

The most important design criteria associated with OO languages, including Python, is the ability to both modify and reuse existing software modules. Doing both of these functions without upsetting or "breaking" the existing code base is the subject of the next section regarding interfaces.

Using Interfaces

If you consider the previous Robot model, you should realize that there are two behaviors that may vary for every robot being modeled. These behaviors are movement and sensing. It would therefore be wise to break out or abstract both these behaviors in such a way as to be able to add new robot models without disturbing the existing code. Using a special class known as an interface is the key to achieving this goal. Figure 1-7 is a UML diagram showing both MoveBehavior and SenseBehavior as interface classes along with appropriate child classes, which implement specific behavior associated with a robot subclass, that is, MoveWith-Wheels for the WheeledRobot and MoveWithTracks for the TrackedRobot. Note that Python makes no distinction between interface classes and normal classes, which is not true for languages such as Java. In any case, it really makes little to no difference, for what is being discussed here is the overall software design approach.

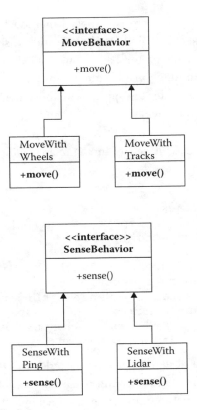

Figure 1-7 *Interface UML diagram.*

The following is the MoveBehavior class definition code listing. It is quite brief and only exists to serve as a parent for the class interface hierarchy.

```
class MoveBehavior():
    def __init__(self):
        pass
```

The SenseBehavior class definition is almost identical to the MoveBehavior class and will not be shown.

Next, the child class definition associated with movement for a wheeled robot is listed.

```
from move_behavior import MoveBehavior
class MoveWithWheels(MoveBehavior):
    def move(self):
        print "I move using wheels"
```

Again, this code listing is quite brief, as the only action that will be performed is a simple print message. In a real-world situation, this class definition could be very complex, involving many commands and physical interfaces associated with a real wheeled robot. The MoveWithTracks class definition is almost identical to the previous listing, except it contains a slightly different print message.

The SenseWithPing class definition is shown next.

```
from sense_behavior import SenseBehavior
class SenseWithPing(SenseBehavior):
    def sense(self):
        print "I use a Ping sensor to determine distance"
```

This code listing follows the same template as the movement behavior child classes, which is a good indication of the simplicity and consistency of this design approach.

The next section ties the Robot class design and the interface design together and presents a test class to prove that the overall project functions as expected.

Integrated Robot Project Design and Test

The complete robot project UML diagram with the robot parent and subclasses, as well as the behavior interface classes, along with the respective subclasses are shown in Figure 1-8. A test class is also shown in the diagram, which is labeled as a client to conform to typical UML designations.

Note that the interface subclasses that represent the behavior portion of the project for specific robots are connected to the appropriate Robot subclass blocks

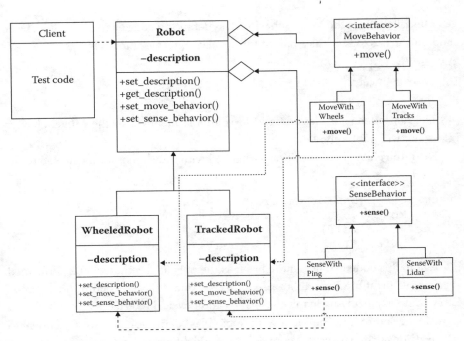

Figure 1-8   *Complete robot project UML diagram.*

using dashed lines. This UML designation indicates that the behavior interface classes help compose part of the class but are not strictly part of the class definition, which is consistent with my earlier discussion regarding composition.

The test code block, labeled client, is also attached to the Robot class with a dashed line, but this relationship is just to indicate that the client needs to instantiate or create some Robot objects to carry out its intended functions. The test code contains no class definitions and simply calls several public methods made available by the Robot class.

The following is the program listing for test_code.py:

```
from  robot import Robot
from  wheeled_robot import WheeledRobot
from  tracked_robot import TrackedRobot
from  move_behavior import MoveBehavior
from  sense_behavior import SenseBehavior
from  move_with_wheels import MoveWithWheels
from  move_with_tracks import MoveWithTracks
from  sense_with_ping import SenseWithPing
from  sense_with_lidar import SenseWithLidar
```

```
if __name__ == "__main__":
    print
    print '_' * 48
    print
    print "Wheeled Robot"
    print
    wheeled = WheeledRobot()
    wheeled.perform_move()
    wheeled.perform_sense()
    print
    print '_' * 48
    print
    print "Tracked Robot"
    print
    tracked = TrackedRobot()
    tracked.perform_move()
    tracked.perform_sense()
    print
    print '_'* 48
    print
```

There are obviously a lot of import statements, as this is the way Python assembles or references all the different classes that make up this project. You should also note that I only instantiated two objects in the code, wheeled and tracked, that reference the WheeledRobot and TrackedRobot classes, respectively. Additionally, all the print statements in the code simply make the program output easier to read.

The output from this program is shown in Figure 1-9 with the four print statements that represent wheeled and tracked robot movement and sense behaviors.

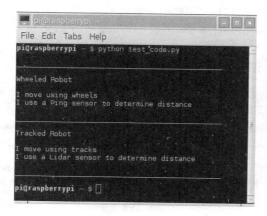

Figure 1-9 *Test code output.*

The previous discussion was lengthy but necessary to illustrate a good approach to creating code that is understandable and fairly easy to adjust to accommodate changing requirements. Another important principle underpinning this methodology is to:

Favor composition over inheritance.

Using the interfaces with the child implementation classes is a concrete example of applying this principle. The code was made robust by recognizing that movement and sense behaviors will be different for different robot types. So instead of trying to hard-code specific types of behavior in each of the concrete Robot subclasses, I chose to pull that all out and place it in the interfaces. This apparent recognition of object differences leads to one more principle:

Identify what varies and encapsulate it.

Movement and sense vary for each type of robot, so those behaviors were pulled out of the Robot class and delegated to interfaces. Then, appropriate interface implementation subclasses were created to invoke the specific behavior for that Robot object with that type of behavior. Very simple and somewhat elegant!

I have saved the best for last in this discussion. The approach I have taken with creating this code is an example of the design pattern strategy. Design patterns (DPs) are often introduced into advanced computer science courses; however, I believe if you learn the patterns early on, you will develop good code design practices that will help you in all your future development efforts. DPs in software development[1] have been around since 1995, when the book *Design Patterns*[1] was published. DPs are not specific solutions to specific problems, but are a methodical approach for creating good solutions given a general type of problem domain. They are based upon years of great software development by masters in this field, "standing on the shoulders of giants," as the old saying goes.

Dynamic Binding

In the Robot class definition two additional methods were added that have not yet been discussed. These are `set_move_behavior(self, mb)` and `set_sense_behavior(self, sb)`. These methods allow Robot objects to reference both movement and sense behaviors as needed during program execution.

---

[1] Erich Gamma, John Vlissides, Ralph Johnson, and Richard Helm, *Design Patterns: Elements of Reusable Object-Oriented*, Addison-Wesley, Boston, MA, 1995.

This modus operandi is called dynamic or late binding, as compared to establishing fixed references that occur during the compilation stage, which is called static or early binding. Dynamic binding is interesting from the perspective that robots can change their movement and sense behaviors to suit real-time conditions. I have created two more interface implementation classes to illustrate dynamic binding. Don't get too excited, however, as all they do is output a slightly different console message compared to the original class message. The two new classes created are StartRapidMoveOnTheFly and StartExtendedSenseOnTheFly. These two new classes are listed next.

```
from move_behavior import MoveBehavior
    class StartRapidMoveOnTheFly(MoveBehavior):
        def move(self):
            print "I am now moving rapidly"

from sense_behavior import SenseBehavior
    class StartExtendedSenseOnTheFly(SenseBehavior):
        def sense(self):
            print "I am now using extended sensing"
```

I also had to revise the test code to demonstrate that dynamic binding took place. The revised test code, which is named test_code_dynamic.py, is shown next.

```
from   robot import Robot
from   wheeled_robot import WheeledRobot
from   tracked_robot import TrackedRobot
from   move_behavior import MoveBehavior
from   sense_behavior import SenseBehavior
from   move_with_wheels import MoveWithWheels
from   move_with_tracks import MoveWithTracks
from   sense_with_ping import SenseWithPing
from   sense_with_lidar import SenseWithLidar
from   start_rapid_move_on_the_fly import StartRapidMoveOnTheFly
from   start_extended_sense_on_the_fly import StartExtendedSenseOnTheFly
if __name__ == "__main__":
    print
    print '_' * 48
    print
    print "Wheeled Robot"
    print
    wheeled = WheeledRobot()
    wheeled.perform_move()
    wheeled.perform_sense()
    print "Now dynamically changing the move"
    move_instance = StartRapidMoveOnTheFly()
    wheeled.set_move_behavior(move_instance)
    wheeled.perform_move()
```

```
print
print '_' * 48
print
print "Tracked Robot"
print
tracked = TrackedRobot()
tracked.perform_move()
tracked.perform_sense()
print "Now dynamically changing the sense"
sense_instance = StartExtendedSenseOnTheFly()
tracked.set_sense_behavior(sense_instance)
tracked.perform_sense()
print
print '_'* 48
print
```

The corresponding terminal output is shown in Figure 1-10.

An additional text message, "Now dynamically changing the move (sense)," is displayed just before the call to perform the move or sense associated with the newly reallocated behavior instance variables move_instance and sense_instance. The messages "I am moving rapidly" and "I am now using extended sensing" come from the new classes I just incorporated into the project. Dynamic binding is also considered polymorphic behavior in that sending exactly the same message invokes different responses from the target objects. In this case, the message (or method call) is `wheeled.perform_move()` for the WheeledRobot object and `tracked.perform_sense()` for the TrackedRobot object.

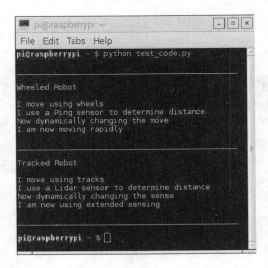

Figure 1-10 *Terminal output with dynamic behavior set.*

On-the-fly or dynamic reconfiguration of robot behavior provides a huge amount of flexibility in coping with real-time situations. Consider the case where a reconnaissance robot has been deployed into a battlefield situation. Sometimes field conditions change, such as the addition of smoke or fog, inhibiting normal robot movement and/or sensing. Reconfiguring the robot to accommodate to the new environment would allow the reconnaissance robot to continue its mission. Installing new behavior to match the mission is easily accomplished with dynamic binding.

It is important to realize that dynamic binding is simply not possible using a strict inheritance structure. The added flexibility that interfaces provide is yet another reason that most developers favor using interfaces over strict inheritance.

This last section concludes my introductory discussions regarding OO, Python basics, and a recommended software design approach. It is now time to start using MicroPython.

Using MicroPython with a Pyboard

The simplest and quickest way to start using MicroPython is to acquire a Pyboard and connect it to a computer using a micro USB cable. I used a version 1.1 Pyboard, which I purchased from Adafruit.com and whose part number is 2390. This Pyboard is shown in Figure 1-11.

Figure 1-11 *Version 1.1 Pyboard.*

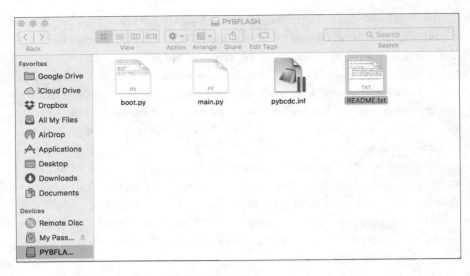

Figure 1-12   *Contents of the PYBFLASH drive.*

The Pyboard is powered through the USB cable and needs no other power supply connection. The computer you use to connect to the Pyboard should have some type of terminal program running in order to establish a communication session. In my case, I used a MacBook Pro with its built-in terminal program. A USB drive named PYBFLASH appeared on the MacBook desktop when I plugged the USB cable into it. Figure 1-12 shows the contents of this drive.

There is a file named README.txt, which you should open, as it contains instructions on how to start a terminal session with the Pyboard according to which OS you are using, that is, Windows, Mac, or Linux. Figure 1-13 shows the contents of this Readme file.

Figure 1-13   *Readme file for establishing a terminal session.*

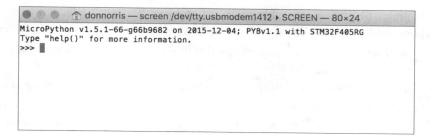

```
● ● ●      ⌂ donnorris — screen /dev/tty.usbmodem1412 ▸ SCREEN — 80×24
MicroPython v1.5.1-66-g66b9682 on 2015-12-04; PYBv1.1 with STM32F405RG
Type "help()" for more information.
>>> ▊
```

Figure 1-14   *MicroPython initial welcome screen.*

I followed the instructions for a Mac and was soon greeted with the opening screen shown in Figure 1-14.

At this point you are interacting with a "regular" Python session such that you can enter expressions and see immediate results as previously shown in the chapter. MicroPython behaves identically to ordinary Python at this level. This interactive interpreter mode is also known as REPL, which is short for read-eval-print-loop. It is one of three ways the user can interact with the Pyboard. The other two modes will be discussed in the next chapter where I explain how to use programs with the board. Meanwhile, the REPL mode is by far the easiest and most convenient way to work with the Pyboard.

You should also be aware that there is only about 200 kB of random access memory (RAM) available, and you will rapidly run out of memory if you attempt to deal with objects and/or expressions that consume substantial memory. However, there is a good option available to help with this constrained memory. That option consists of a micro SD card socket, which you can see is located next to the micro USB connector in Figure 1-11. Plugging a 4-GB card into this socket will vastly enhance the amount of memory available to the Python OS and allow you to create basically unlimited sized programs. My strong recommendation is that you only use a relatively high-speed class 10 micro SD card in order to minimize any processing delays that will likely occur when transferring data between the card and the Pyboard.

Bare-Metal Approach

I am concluding this first chapter with a discussion of the bare-metal approach that Dr. George decided to take when creating MicroPython. These points come in large part from his July 17, 2015, podcast "Python on bare metal with MicroPython" on talkpython.fm, episode #17.

It should be made clear at this point that MicroPython is a complete rewrite of Python version 3.4. The main goals in this rewrite were to make this Python version as efficient as possible and to use the least amount of RAM necessary to execute any operation.

The phrase "bare metal" in the section heading refers to the fact that MicroPython takes on the role of an OS for the microcontroller in which it is installed. No other OS is required or needed, and MicroPython has complete access to and control over the microcontroller's machine code—hence the bare metal reference. Having so much low-level control allows MicroPython to initiate events such as disabling hardware interrupts such that a program can effectively use direct machine cycle counts without being interrupted. This level of control is simply not possible in ordinary Python microcontroller installations, such as may be found in the RasPi I used earlier in the chapter. Having such intimate control of machine operations allows the developer to create the most efficient programs possible, while still programming at a relatively high abstract level with the Python language.

It is also possible to include in-line assembly code into a MicroPython program. You should also be aware that MicroPython has an option to generate assembly code in any one of three formats: bytecode, ARM code, and finally direct machine code. The bytecode format is the default mode for the MicroPython compiler. Every instruction or data entry contained in a Python program is eventually compiled into bytecode before being executed. Some developers seem to be ignorant of this fact when they state that Python is slow. It is a strictly interpreted language as compared to a strictly compiled language such as C. The truth is that although Python does use a line-by-line interpreter, it also employs a very fast compiler for bytecode generation. This approach also mirrors how the Java language functions, which is good. CPython, which is based on a C runtime, actually is comparable in terms of speed with MicroPython running on a microcontroller such as the Pyboard. However, this comparison is somewhat faulty and is really like comparing apples to oranges in that you cannot load CPython onto a Pyboard due to memory constraints. Speed comparisons actually can be made on a PC, where it is totally possible to run instances of both MicroPython and CPython. However, the key point to realize here is that MicroPython is no longer running in a bare-metal environment and must operate in compliance with a host OS, which certainly hinders its efficiency. Nonetheless, it still operates quite well.

Dr. George also points out that MicroPython was never meant to replace C, but just to provide developers with an extremely efficient way to program a microcontroller while using the friendly, high-level Python language.

This last section concludes this chapter's introductory discussions. Much more detailed information on MicroPython starts in the next chapter.

Summary

The chapter began with a brief history of the MicroPython project and its connection to the regular Python language.

A discussion of object-oriented concepts followed as I considered it important that readers understand these principles in order to fully appreciate and properly utilize this language.

The next section addressed how to model a robot using OO techniques. I also briefly discussed some basic Python concepts in order to lay the groundwork for the Python code that followed. This code implemented a robot parent class and several specific robot child classes.

I next introduced the concept of interfaces and explained how they can greatly improve the quality and maintainability of a code project. These interfaces were then incorporated into a comprehensive robot package that included some test code to demonstrate that the code functioned as desired. The test code was then modified slightly to demonstrate how dynamic binding worked.

The next section demonstrated how to boot up MicroPython on a Pyboard using the REPL mode of operation.

The final section was a brief description of how MicroPython functions as an OS in a bare-metal manner on the Pyboard. There were also some thoughts and insights on MicroPython from its creator, Dr. Damien George.

2

Introduction to the Pyboard

This chapter will focus on how MicroPython is used with the Pyboard. It is not intended to be a reference chapter, but more of personal introduction where I explain the hardware and how the MicroPython software enables the board functions. This approach seems to me a good one in that MicroPython is crafted to be very "close" to the way the hardware functions. However, feel free to go back and reference this material as you progress through the remainder of the book, as you will likely need to refresh yourself on certain topics that may seem cloudy or unclear the first time you encounter them.

Pyboard Hardware

The Pyboard was especially created for MicroPython. It is not a repurposing or porting of an existing microcontroller board. It is also completely open source, with schematics freely available for your own use, whether for personal or commercial use, under the terms of the MIT license. I do urge you to review this license before you attempt to commercialize a product based on the Pyboard design. The license is extremely generous in nature, but you do have to comply with some conditions, especially recognizing the design's creator.

I have included a Pyboard schematic in Figure 2-1. The physical Pyboard was shown in the previous chapter.

I realize that this figure is probably quite difficult to interpret given the book's limited dimensions. You download the schematic in a PDF format from the www .micropython.org website. It can be found in the Docs area.

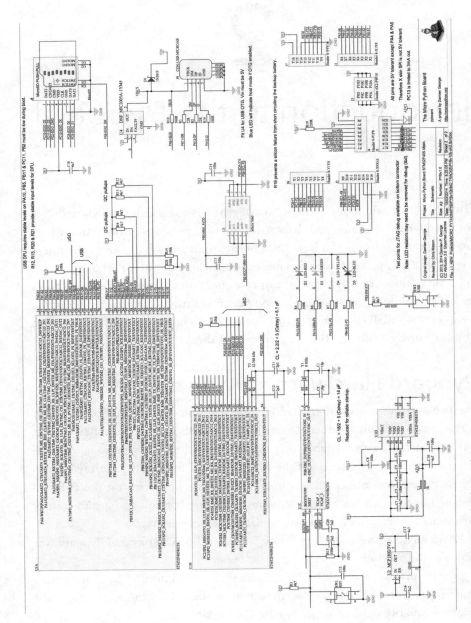

Figure 2-1 Pyboard schematic.

Schematic Designation	Manufacturer and Part Number	Description
U1	STM semiconductor STM32F405RGT6	32-bit microcontroller.
U2	Microchip MCP1802/3V3	Low-dropout voltage regulator. Converts 5 V from USB to 3.3 V for U1.
U3	Freescale semiconductor MMA7660	Three-axes accelerometer with I2C output
U4	Microchip MIC2005A	USB power switch. Only required if USB OTG is enabled.

Table 2-1 *Pyboard Integrated Chips and Their Functions*

The reason I elected to show you this schematic is to point out that there are only four integrated chips on the board. Table 2-1 lists these chips and their principal functions.

The main chip—and the largest one—is the STM microcontroller chip (known as STM micro henceforth), which is the heart of the Pyboard. It is a sophisticated 32-bit microcontroller with an ARM Cortex-M4 processor core. Some of the key specifications for this chip are detailed in Table 2-2. It is important to have a good understanding of this chip because the hardware and software are so intertwined.

A complete PDF datasheet for this STM microcontroller can be downloaded from the STM website at www.st.com. I highly recommend that you download this 201-page PDF, as it contains detailed explanations of all the microcontroller features you will need when creating programs for this chip. Based upon my experience as a professional embedded developer, having a microcontroller datasheet readily available is an invaluable asset for successful program development. There are just too many fine details regarding modern-day microcontrollers, which makes it imperative to read and understand the information contained in the datasheet, which has been prepared by the engineering design team.

At this point, I will digress a bit and discuss some key aspects of generic microcontroller programs, independent of any implementation language such as MicroPython.

Item	Quantity	Remarks
Flash ROM	1024 kB	
RAM	192 kB	
Ethernet	none	
General-purpose timer	14	Numbered 1 to 14; three are reserved for internal use
Random noise generator	1	RNG
Serial peripheral interface	3	SPI, a bit serial interface
Universal synchronous asynchronous receiver transmitter	4	USART and UART, bit serial interfaces
Inter-integrated circuit	3	I2C, a bit serial interface
USB	1	USB and USB/OTG
Controller area network	2	CAN, a bit serial interface
Secure digital input/output	1	SDIO, enables the SD card
General-purpose input/output pins	51	GPIO
Analog-to-digital converter	16	12-bit ADC
Digital-to-analog converter	2	12-bit DAC
Clock speed	168 MHz max	8 MHz xtal version 1.0 12 MHz xtal version 1.1 PLL for about 10× speedup
Real-time clock	32.768 kHz	RTC
Power supply voltage	1.8 to 3.6V	Set at 3.3V using U2
Package	LQFP64	

Table 2-2 *Key Specifications for the Pyboard STM Micro*

Generic Microcontroller Program Development

In the late 1970s, an extremely popular 8-bit microcontroller named the 8051 was introduced by Intel. It was quite limited in its capabilities and functionalities as compared with today's 32-bit microcontrollers, but later versions are still incorporated into many millions of products and projects. The interesting fact about the 8051 is that there is no host operating system for this chip. Any program executed on an 8051 was required to implement all memory, file, and resource utilities as needed and at the same time keep program sizes to an absolute minimum. Fortunately for most developers, an excellent set of C libraries were available that met most of the project requirements, yet used precious little of the available ROM memory space. There was essentially only a single program run on the chip, which consisted of an infinite loop that had to be interrupted at times to service events such as data entry or taking a sensor measurement. This process of using interrupts became second nature to not only 8051 developers, but to anyone else using a similarly configured chip. This situation still holds true in today's development world where interrupts are still critical for efficient and effective microcontroller operations. The STM micro used on the Pyboard offers a whole array of different interrupts to help create programs that are both efficient and responsive. The MicroPython language offers high-level, direct control over interrupts, which is a unique feature not found in any other object-oriented (OO) language of which I am aware.

The preceding brief discussion was a prelude and justification for this next section, which discusses interrupts and where I will also create a simple interrupt demonstration using MicroPython.

Interrupts

Interrupts generally have three sources, as shown in Figure 2-2. I will discuss each source separately and provide a framework by which you can determine when and

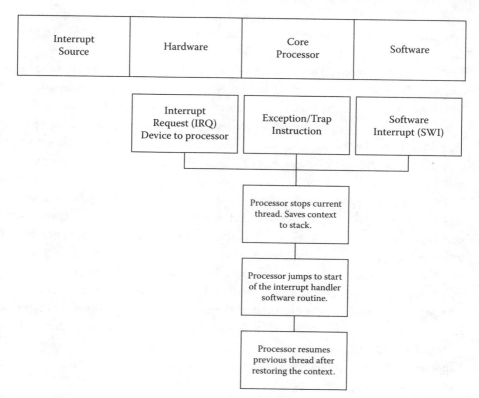

Figure 2-2 *Interrupt sources and flow diagram.*

where to use a particular type of interrupt. The types of interrupt sources include the following:

- Hardware
- Exceptions and traps
- Direct software instructions

Hardware interrupts may be generated by a vast array of devices that need the attention of the processor. Interrupt events might be the immediate availability of sensor data or the sudden increase in a device input voltage level over some preset threshold value. These hardware interrupt requests (IRQs) are directly connected to the STM micro using any one of 16 GPIO lines. The STM micro also has a built-in interrupt/event controller (EXTI) to manage all of these possible IRQs in an orderly fashion.

Exceptions and traps refers to cases where either an abnormal event happens during program execution (exception) or the program purposely branches to an interrupt routine based on a preset condition or state (trap). These types of interrupts may also be broadly categorized as a specialized software interrupt; however, they are primarily focused on overall events that may or may not happen during program execution.

Forced or deterministic software interrupts (SWIs) are the last interrupt source category. SWIs are places in a program where an interrupt is forced for a particular reason. It often may be used to terminate a program at a specific time or after a preset computed value has been reached. An SWI could also be used to jump to an interrupt handler based upon some metric regarding the amount of memory used so that additional memory can be allocated before the program is abnormally terminated through an out-of-memory exception.

Figure 2-2 also shows a generalized diagram of what steps the processor will take when an interrupt is recognized and processed. The first step is always to complete processing the current instruction that was being handled when the interrupt happened. Next the processor state is saved on to a memory stack, which is just a series of preset memory locations where data such as the program counter (pc) is stored. The pc is important, as that contains the memory location for the next instruction to be executed by the processor. This shows the processor "knows" where to resume its normal processor flow after the interrupt is handled. It simply jumps back to the restored pc location and begins again as if nothing had happened. Of course, other key data registers are also automatically restored so that the complete processor context is as it was prior to the interrupt.

The processor will jump to a preset interrupt handler routine based on the interrupt source and its priority. The STM micro has another built-in controller, named the nested vectored interrupt controller (NVIC), which handles up to 16 priority levels and up to 82 maskable interrupt channels, plus the 16 interrupt lines of the STM micro. Maskable interrupts are those that can be software disabled while still having some higher-priority nonmaskable interrupts activated. Of course, disabling all interrupts prevents both maskable and nonmaskable interrupts from interrupting the processor. The NVIC knows the address of the interrupt handler to jump to based on the interrupt source and its priority. The term *nested* in the NVIC name refers to the fact that it allows higher-priority interrupts to interrupt lower-priority ones while keeping track of all the return addresses so that the processor does not fail in processing all these interrupts in a proper sequence.

Hardware interrupts from connected devices on the STM micro can be triggered by a variety of changes occurring on the pin connection. These include

1. Voltage transition from low to high (rising edge trigger)
2. Voltage transition from high to low (trailing edge trigger)
3. Voltage state change from low to high
4. Voltage state change from high to low

There is only a slight timing difference between the edge triggers and state change triggers, where edge triggers are more sensitive to an immediate IRQ, whereas the state change has a small settling time before triggering an IRQ. Select the appropriate type that best matches the timing and interrupt requirements for your project.

My final point in this discussion is to point out that each interrupt handler must be designed to completely satisfy the requirements of the device or software event that triggered the interrupt event. Failing to do so may cause the overall project design to fall short of meeting all its design requirements.

Before demonstrating a program that implements some real interrupts, I need to cover some basic MicroPython functions that control the Pyboard's onboard light-emitting diodes (LEDs).

Controlling the Pyboard

Prior to doing any of the projects, I would strongly recommend that you solder pin sockets to the Pyboard as shown in Figure 2-3.

Figure 2-3 *Pyboard with a fully populated set of pin sockets.*

Having pin sockets available makes the process of connecting the Pyboard to a solderless breadboard quite easy and facilitates the setup of the book's projects. Additionally, I used a set of inexpensive jumper wires to make the connections between components on the breadboard and the Pyboard. You can also use ordinary hookup wire, but I have found the latter technique to be troublesome, with occasional poor connections, compared with using manufactured jumpers.

I will first start the hardware demonstrations with the built-in LEDs. The Pyboard has four programmable colored LEDs as listed in Table 2-3.

Enter the following command at the read-eval-print-loop (REPL) prompt to turn on the red LED:

```
pyb.LED(1).on()
```

Pyboard ID	LED Color	MicroPython Reference #
D2	Red	1
D3	Green	2
D4	Yellow	3
D5	Blue	4

Table 2-3 *Pyboard LEDs*

The command to turn off the LED is similar:

```
pyb.LED(1).off()
```

Simply change the number in the LED argument parentheses to control a different-colored LED according to the reference number shown in Table 2-3.

Next I will show how to use an external push button with an external LED. The object of this demonstration is simply to light the LED whenever the button is pressed.

CAUTION *The maximum current that maybe sinked or sourced from any single GPIO or control pin on the Pyboard is 25 mA. The overall total current that may be sinked or sourced from the Pyboard is 240 mA. This means you can never have more than nine pins simultaneously drawing maximum current. Furthermore, you should always use a 220-Ω or greater current-limiting resistor in series with a device such as an LED. Figure 2-4 shows the circuit, including both the push button input and LED output.*

Note that the X1 GPIO push button input is pulled up with an internal 40-KΩ resistor, which is then pulled low when the push button is pressed. Detecting the button press thus becomes a matter of sensing a high-to-low voltage transition on the GPIO pin dedicated to this input. Using this type of input configuration minimizes any possible false sensing due to a "floating" voltage transient occurring on the open-circuit high-impedance input. Using this method for push button signaling is generally the preferred design approach to make a project more robust and reliable.

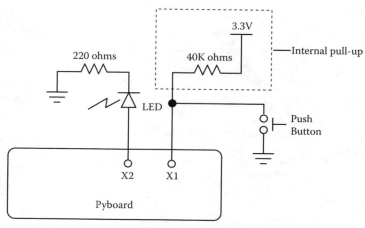

Figure 2-4 *Schematic for push button/LED demonstration.*

Figure 2-5   *Physical setup for the demonstration circuit.*

The X2 GPIO pin output controls a typical LED in series with a 220-Ω current-limiting resistor. Figure 2-5 shows the physical configuration for the demonstration circuit.

The test software is in the form of a program (Python script) that is added to the main.py file found in the PYBFLASH USB drive, which is automatically created when the Pyboard is plugged into the computer. You can use any suitable text editor to modify the main.py file. I happened to use the XCode editor on my MacBook Pro, as it was already configured to handle these file types. In Windows, you could use either Notepad or Wordpad if they are available on your system. Linux users can use vi or nano for this task. Just remember to disable any special formatting in the text editor because those characters will definitely interfere with creating a working program.

Python Test Program

The following is the program that was added to the main.py file, which is located in the PYBFLASH drive:

```
#PushButton.py added to main.py in the PYBFLASH drive
#Created by D.J. Norris 2/2016
from pyb import Pin
pinX1 = Pin('X1', Pin.IN, Pin.PULL_UP)
pinX2 = Pin('X2', Pin.OUT_PP)
```

```
while True:
    if pinX1.value() == 0:
        pinX2.high()
    else:
        pinX2.low()
```

The pinX1.value() == 0 statement continually checks for a low value on pin X1 and will set pin X2 to high when it senses a low value; otherwise, it sets pin X2 to low. This software design is known as polling and does demand continuous STM microprocessor cycles. It is not as efficient as using an interrupt design, which will be demonstrated later in the chapter, but on the flip side, it is a simple software design and easy to implement.

The LED did light for as long as the push button was pressed, which confirmed that the hardware and software functioned as desired.

I would also like to comment on the program code to point out some OO concepts that appear in this program and were first discussed in the previous chapter. pinX1 and pinX2 are two objects instantiated from the Pin class for which various methods are called to carry out the necessary program functions. For instance, the statement pinX2.high() sets the X2 pin to a high or 3.3-V state. The high() portion of the statement is the method call, and the "." between the object name pinX2 and the method call is the "member of" operator. This is how Python calls or invokes a specific method, which is part of the Pin class definition.

The from pyb import Pin statement directs Python to use the Pin class, which is part of the pyb library. The pyb library contains many more class definitions that I will be using throughout this book, and it is certainly the single most import asset for making MicroPython a functioning language.

The next demonstration program simply uses the built-in LEDs and does not require any additional components. It does start to show some ways timing can be utilized in a MicroPython program.

Blinking LEDs

Two versions of a program will be used in this section: one that uses timing only available in MicroPython and the other that uses a regular Python timing function. My purpose in this exercise is to show you that MicroPython offers some unique features for efficient timing operation not available in ordinary Python.

PyBlink

The first program is the one based on using the Python time library. You should note that the values used in the sleep method argument are in seconds. I named the program PyBlink and added it to the main.py file as previously discussed. The file is listed next:

```
import time
import pyb
green = 2
yellow = 3

while True:
    pyb.LED(green).on()
    pyb.LED(yellow).on()
    time.sleep(1)
    pyb.LED(green).off()
    pyb.LED(yellow).off()
    time.sleep(1)
```

Please note that I used not only the time library but also the pyb library. The latter was used to directly call the on and off methods for the green and yellow LEDs. You should compare this approach to the way I performed a similar operation in the previous program. In the PushButton program, I had to instantiate an object that represented or modeled a GPIO pin configured as an output, which in turn controlled an LED. In this example, the LEDs are statically identified by their fixed reference numbers and do not need an explicit object instantiated in order to call a method. This situation is only made possible because the LEDs are "hardwired" to the STM micro and cannot be reconfigured, which is normally the case for GPIO pins.

I added this program to main.py and executed it by simply connecting the Pyboard to the laptop. Both the green and yellow LEDs blinked at a one-second rate as expected.

One other comment regarding the program, which addresses a common mistake that developers often make. That is the use of "magic numbers" within an expression or function. In this program, a magic number would be 2 if used in the expression for controlling the green LED. Anyone looking at this program without prior knowledge that 2 was the predefined reference number for the green LED would have wondered about the meaning of the number in the expression— hence, the use of the word *magic*. In order to avoid this situation, I defined the word green as 2, and the language substituted that value for green wherever it was used in an expression. This makes it clear that the green LED is being controlled, thus eliminating the use of a magic number. I highly recommend this design approach, which should make your programs much more understandable and maintainable.

PyBlink_MP

The second version of the LED-blinking program is named PyBlink_MP, where the MP suffix indicates that it is especially crafted to use some MicroPython features. The program is listed next with comments inserted to indicate where it differs from the previous version.

```
#no time class import required
import pyb
green = 2
yellow = 3

while True:
    pyb.LED(green).on()
    pyb.LED(yellow).on()
    #the delay method in pyb uses milliseconds for units of delay
    pyb.delay(1000)
    pyb.LED(green).off()
    pyb.LED(yellow).off()
    pyb.delay(1000)
```

Again, when this code executed, I observed no differences between the previous blink behavior and the blink behavior controlled by this program. The real difference between these two programs is a bit subtle, as each of the delay methods takes a slightly different approach to implementing a specific time delay. The differences are not applicable to a Pyboard because MicroPython is also the host OS. The sleep method used in the time class, when run on a regular pc, is dependent on the host system clock ticks. These ticks may not come on a regular basis or specific schedule to the running Python program due to inherent host scheduling, where the host OS might have to attend to higher-priority tasks before feeding clock ticks back to the Python program. This redirection can cause uncertainty and poor clock tick granularity compared with MicroPython, where the currently executing program has the undivided attention of the MicroPython OS. In other words, MicroPython on the Pyboard fits the role of a real-time OS much more than ordinary Python or MicroPython running on a normal pc. I am really not suggesting that the Pyboard running MicroPython can be considered a true real-time system, but it is much closer to fulfilling that role compared with an ordinary pc running Python. Of course, all this discussion regarding real-time performance becomes somewhat moot when interrupts are considered, which I will discuss in the next section.

Hardware Interrupt Demonstration

I will start this section by stating that much of this material is based on the fine tutorials available on the www.micropython.org website. This first interrupt demonstration involves using one of the built-in Pyboard switches. Two push button switches are installed on the board. One is labeled RST, which, when pushed, will cause a hard reset of the Pyboard. This is equivalent to switching the power off and then on again to the board. This reset will also abruptly disconnect the PYB-FLASH USB drive, which could cause some problems compared with properly ejecting the drive before powering off.

The second push button is labeled USR—this is short for User—and is an uncommitted, general-purpose switch. All you need to do is instantiate a Switch object using the following command:

```
sw = pyb.Switch()
```

Of course, you will need to import the pyb library; otherwise, MicroPython will complain that it does not exist.

Next, simply enter sw() at the REPL prompt to display the USR switch state, that is, False for not pressed and True for pressed. Try it out for yourself.

Although the switch is a very simple device, along with its Switch class, it does have a very interesting function associated with it. This function is known as a callback function because it sets up an expression and/or snippet of code to be executed whenever the switch is pressed. The easiest way to demonstrate this callback function is with a simple example. Enter the following at the REPL prompt:

```
sw.callback(lambda:print("switch pressed!"))
```

Next, press the USR switch and observe the print statement displayed after the REPL prompt. Note, you will have to press the ENTER key to get the REPL prompt to show once again. This command sets up the Switch class callback function to interrupt the REPL loop and display the test message when the USR button is pressed.

On an aside, notice that the word *lambda* was used in the previous command. Lambda is a Python key word used to identify an anonymous function, which is a function that is dynamically created with no unique name. This is a very handy feature, which is also found in a variety of OO languages.

I next expanded this series of interrupt demonstrations by writing a program that continually blinked the green LED but could be interrupted by pressing the USR switch. Then the yellow LED would blink for five seconds, at which point it would cease blinking and the green LED would continue to blink. Also note that in the program, which I named BlinkInterrupt, is an interrupt service routine (ISR) function explicitly defined as BlinkYellow. This shows that it is possible to use not only an anonymous function but also explicit routines as interrupt handlers, or as they are more commonly known, ISRs.

```python
#BlinkInterrupt
#Created by D. J. Norris 2/2016
import pyb

#Avoid "magic" numbers
green = 2
yellow = 3

#Interrupt service routine (ISR)
def BlinkYellow():
    for x in range(0,4):
        pyb.LED(yellow).on()
        pyb.delay(1000)
        pyb.LED(yellow).off()
        pyb.delay(1000)

sw = pyb.Switch()
sw.callback(BlinkYellow)

#Forever loop
while True:
    pyb.LED(green).on()
    pyb.delay(1000)
    pyb.LED(green).off()
    pyb.delay(1000)
```

The green LED immediately began to blink after the Pyboard was plugged into the laptop. It ceased blinking when the USR button was pressed, which then caused the yellow LED to blink for five seconds. The green LED did not blink during this interval because the forever loop was interrupted by the ISR. It did resume blinking after the ISR completed its operations.

I need to make a few observations regarding ISRs. First, they should be fairly short and relatively uncomplicated because their purpose is to simply address handling the interrupt source and nothing else. Any complicated processing should be deferred to the main program after the ISR completes. Data may be passed between an ISR and the main program using global variables, or simply use a flag variable if only a state change notification is required. It is also important to know that ISRs cannot allocate any memory—consequently, ISRs cannot instantiate any Python objects. Obviously, an ISR can use an object previously created in the main program, as was done in this demonstration program.

It is also recommended that a bytearray object be used in the case where multiple data have to be exchanged between an ISR and the main program. In addition, all interrupts should be disabled in the short time interval it takes to fill the array; otherwise, data corruption will likely occur.

Emergency Exception Buffer

An ISR requires a way to log or record an abnormal program situation or exception that might happen while it is running. MicroPython does not transfer any exceptions from the ISR to the main code. The use of an emergency exception buffer fulfills this role quite well. It is nothing more than an allocated memory space where MicroPython can store messages related to the abnormal occurrence that took place while in the ISR. The following statements should be placed in the main code to set up this buffer:

```
import micropython
micropython.alloc_emergency_exception_buf(100)
```

These statements allocate 100 bytes to this buffer, which should be more than adequate to record any execution abnormalities and help assist in debugging.

Now the discussion will shift focus slightly to discuss some of the built-in STM micro timers. Interrupts are still involved in this discussion, as this is the way they interact with the processor.

Timers

A variety of built-in hardware timers are available for your use in the STM micro. These timers are easily made available in the MicroPython language using the Timer class. I will start this discussion using the REPL prompt. Just ensure that the main.py file is empty, which allows MicroPython to boot directly into the REPL prompt.

Enter the following commands at the prompt to create a simple timer object:

```
import pyb
tim = pyb.Timer(2)
```

This new object is completely uninitialized, as you can see if you next enter tim at the prompt. The response is:

```
Timer(2)
```

All this tells you is that the tim reference is associated with timer number 2. This timer can be initialized to cycle or repeat at a 1-Hz rate by entering this:

```
tim.init(freq = 1)
```

Repeating the tim entry now produces:

```
Timer(2, freq=1, prescaler=0, period=83999999, mode=UP, div=1)
```

This data is much more informative, telling us that the timer is running at the peripheral clock (prescaler=0) and it will count up to 83999999 (period=83999999) before resetting and resuming the count up from 0 (mode=UP), which sets up the timer for a 1-Hz rate. The 83999999 period value reflects the 8.4-MHz peripheral clock rate. You may also change the frequency to a much slower rate to see the prescaler kick in. Change the rate to once every 100 seconds by entering this command:

```
tim.init(freq = 0.01)
```

Then entering tim will produce the following:

```
Timer(2, freq=0, prescaler=99, period=83999991, mode=UP, div=1)
```

This display shows that the timer 2 prescaler has divided the peripheral clock rate by 100 even though it shows as 99 because it is 0 based. The period is approximately the same value as for the 1-Hz case, but the overall timer rate is 100 times slower because the scaled input clock is 100 times slower compared with the 1-Hz case.

The preceding demonstrations simply show how MicroPython can instantiate and initialize a timer object. The real power in using timers comes from the fact they will interrupt the main program flow when their preset count has been reached. How this is accomplished is the subject of the next section.

Timer Interrupts

I feel the simplest way to discuss timer interrupts is to show you a simple demonstration program that will blink LEDs using timers instead of delay or sleep statements. The program is named BlinkTimer and is added to main.py as we have done in the past:

```
#BlinkTimer
#adapted by D. J. Norris 2/2016
#Original from "Writing Interrupt Handlers", micropython.org tutorial
import pyb, micropython
micropython.alloc_emergency_exception_buf(100)
green = 2
yellow = 3

class Blink(object):
    def __init__(self, timer, led):
        self.led = led
        timer.callback(self.cb)

    def cb(self, tim):
        self.led.toggle()


greenLED = Blink(pyb.Timer(4, freq=1), pyb.LED(green))
yellowLED = Blink(pyb.Timer(2, freq=2), pyb.LED(yellow))
```

When this program runs you will observe the green LED blinking once per second and the yellow LED blinking two times per second in accordance with the rates assigned in their objects, greenLED and yellowLED, respectively.

In this code, the greenLED instance of the Blink class is composed of a timer 4 object and a green LED object. Timer 4 is set at a 1-Hz count rate, meaning that it will cause an interrupt every second. Once the interrupt occurs, the timer callback function is called and the green LED is toggled. The same operations hold true for the yellowLED object, except this timer is number 2 and set at a 2-Hz rate, meaning it will blink two times per second.

You should notice that you may instantiate as many timer objects as needed up to the limit of the physical timers present in the hardware. In addition, the use of class methods means each timer can share these methods without the need for separate class definitions. This is certainly one of the benefits of using an OO design.

One design feature in this program, which is not too obvious, is that the callback function uses the "self" reference in its function argument. Using this reference will allow every timer object to keep track of its own instance variable(s), thus allowing the object to store data such as the number of times it has invoked the callback function. Storing such data for this program doesn't make much sense, but it would be useful if the Pyboard was controlling turnstiles into a venue, for example. In that case, each callback would represent one paying customer, which would indeed be useful information. Of course, there could be many turnstile objects, each with its own count variable instance.

Other Pyboard Hardware

In this section I decided to discuss two Pyboard hardware features that are not timers, per se, but do depend on the peripheral clock for their operation. These are the analog-to-digital converter (ADC) and the digital-to-analog converter (DAC). I will mainly demonstrate these devices using the REPL mode, except for a DAC program generating a sine wave.

At this point I would highly recommend you download the Pyboard pin-out diagram from www.micropython.org. I show this diagram in Figure 2-6; however, you should download and print out the full color version, as it will be much easier to read and interpret.

I have found this figure to be invaluable in identifying the proper pins for connecting external devices or components. You definitely do not want to mistakenly connect something to an inappropriate pin, as that could permanently damage the Pyboard.

The ADC

The first thing you need to do is import the pyb library because that contains the ADC class, which allows you to instantiate an ADC object. The following instantiation command requires the proper identification of an ADC Pyboard pin, which I show in the commands:

```
import pyb
adc = pyb.ADC(pyb.PIN.board.X1)
```

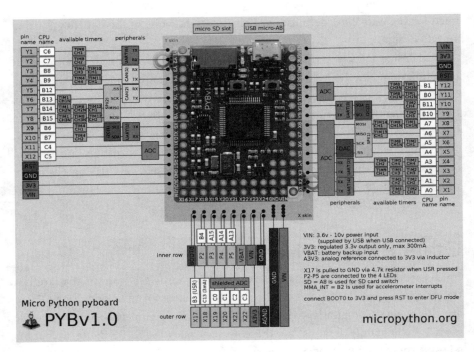

Figure 2-6   *Pyboard pin-out diagram.*

You should next jumper the X1 pin to the 3V3 pin in preparation for taking an ADC reading. I used one of my manufactured jumpers to do this. The jumper connection is shown in Figure 2-7.

Figure 2-7   *X1 pin jumped to the 3V3 pin.*

The command to take a reading with this object is:

```
val = adc.read()
```

Now before I display the reading, I will take a moment to compute the expected reading. All the ADC channels are 12 bits in resolution, meaning that the ADC digital output ranges from 0 to 4095, where 4095 equals $2^{12} - 1$. The X1 ADC input has been jumped to the 3V3 pin, which also happens to be the maximum voltage that can be input into the ADC. Therefore, the expected ADC reading should be 4095. It turns out that when I entered val at the prompt I saw:

```
4095
```

This was precisely the reading I was expecting, which confirmed that the ADC was functioning properly.

I next connected the X1 pin to ground and repeated the ADC read command so that the val variable would be updated. It displayed 0 as was expected.

I will next discuss the DAC, which is the complementary function to the ADC.

The DAC

There are only two DAC channels, which are connected to pins X5 and X6, as you may confirm from Figure 2-6. The DAC class is in the pyb library, and you can instantiate a DAC object by entering the following:

```
from pyb import DAC
dac = DAC(1)
```

You should note that this class definition apparently included two magic numbers: 1 to represent pin X5 and 2 for X6. I really don't approve, but heck, I didn't write the application programming interface (API) for this device. It just reinforces my earlier comments that it is very important to read and understand the documentation supporting the hardware and software. It is important to understand that I am not criticizing the software developers, but simply pointing out that some consistency would be nice when creating the APIs. The ADC class, in contrast to this class definition, requires a full pin definition be inserted when instantiating an ADC object.

The DAC is an 8-bit device, meaning that the object argument takes on values from 0 to 255. The DAC full-scale voltage output is the nominal 3V3 supply, which means an argument value of 255 should produce a 3.3-V output. Of course, the real voltage output is dependent on the actual 3V3 power supply. In my case,

DAC Input Value	Output Voltage (VDC)
0	0.0005
63	0.811
127	1.634
191	2.457
255	3.295

Table 2-4 *DAC Linearity Measurements*

I measured the supply voltage with an uncalibrated volt-ohm meter (VOM) and determined it to be 3.295 V—close, but not precisely 3.300 V. When I entered dac.write(255) at the REPL prompt, I measured 3.295 V at the X5 pin using the same VOM that I used for the power supply measurement. This is precisely the same 3V3 voltage I had earlier measured, which means there is no drop-off in the DAC circuitry.

I next proceeded with a series of measurements to check the DAC's linearity because this is an important operational function. Table 2-4 shows the results of the linearity measurements.

I next charted these measurements because that is a quick way to check the linearity. I expected to see a straight-line relationship, and that was exactly what is shown in Figure 2-8.

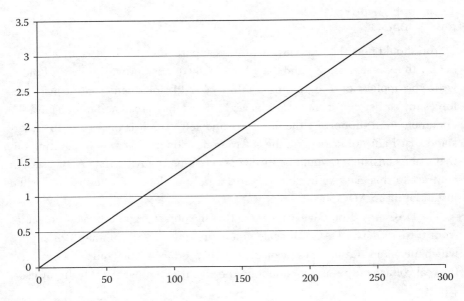

Figure 2-8 *DAC linearity graph.*

I next entered a small program from the micropython.org DAC tutorial to demonstrate how to generate a sine wave. I named this program DAC_Sine.py and list it here:

```
#DAC_Sine.py
#Micropython.org DAC tutorial
# create a buffer containing a sine-wave
buf = bytearray(100)
for i in range(len(buf)):
    buf[i] = 128 + int(127 * math.sin(2 * math.pi * i / len(buf)))

# output the sine-wave at 400Hz
dac = DAC(1)
dac.write_timed(buf, 400 * len(buf), mode=DAC.CIRCULAR)
```

The program continuously generates a 400-Hz sine wave, which is shown in Figure 2-9. It is a screenshot taken from my USB oscilloscope.

The output frequency can be altered by changing the integer multiplier located in the `write_timed ()` method. I tried increasing the number and eventually found that the maximum sine wave frequency generated by the DAC was

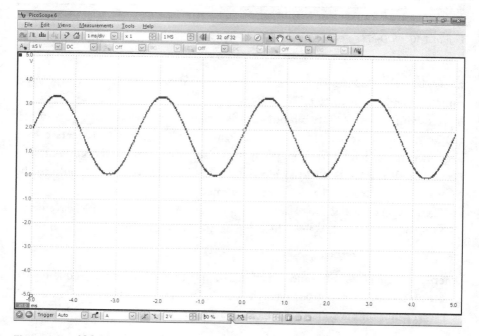

Figure 2-9   *400-Hz sine wave.*

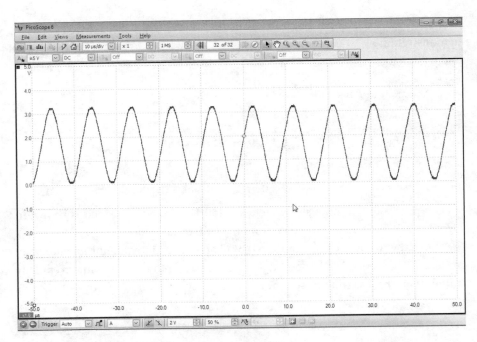

Figure 2-10   *100-kHz sine wave.*

approximately 100 kHz. Figure 2-10 shows this sine wave, and as you can see, it is quite clean and undistorted.

According to the API, two modes are associated with the `write_timed ()` method. The first one, `DAC.CIRCULAR`, was used earlier and repeats a byte sequence at a specified frequency, which is also another method argument. The second mode is `DAC.NORMAL`, which I didn't try, but believe it simply outputs the byte sequence referenced as a method argument.

The DAC class also features two functions that will generate specific waveforms. The first one is a triangle wave, which can be created at the REPL prompt with:

```
dac.triangle(freq)
```

where the actual freq value entered must be 2,048 times the desired frequency. To create an approximate 500-Hz triangle waveform, I entered 1000000. Figure 2-11 is a screenshot of this 500-Hz triangular waveform.

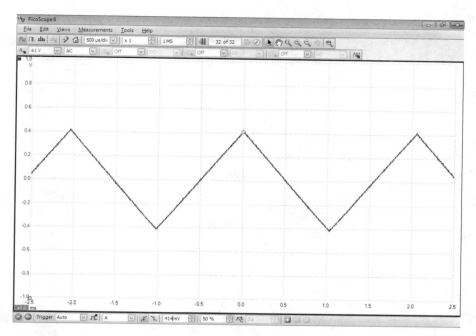

Figure 2-11 *500-Hz triangular wave.*

The next predefined waveform is generated by entering:

```
dac.noise(freq)
```

where freq is a number that apparently represents some pseudo-random noise frequency range. I entered 100000 for a freq number and saw the waveform shown in Figure 2-12 on my USB oscilloscope.

I am certain that this waveform cannot be classified as pseudo-random as it has a very definite waveform and the corresponding spectrum displays well-ordered harmonics, which are very uncharacteristic of a true pseudo-random source. The spectrum for this "noise" waveform is shown in Figure 2-13.

Perhaps there is a use for this particular waveform, but I cannot recommend it as a pseudo-random noise source. What I would recommend as an alternative is to generate a random-number byte sequence of sufficient length and then use the write_timed method in the DAC.CIRCULAR mode to generate a real pseudo-random noise source.

This last section concludes this chapter. The next chapter explores using the extended libraries with MicroPython to greatly increase the Pyboard's capabilities.

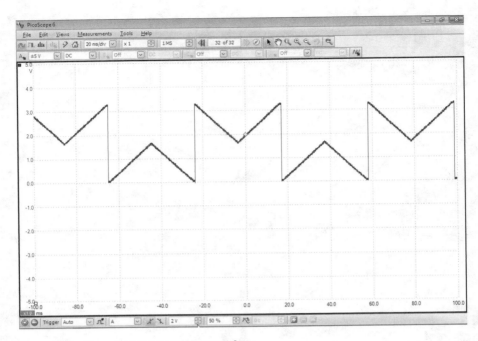

Figure 2-12   *Pseudo-random noise waveform.*

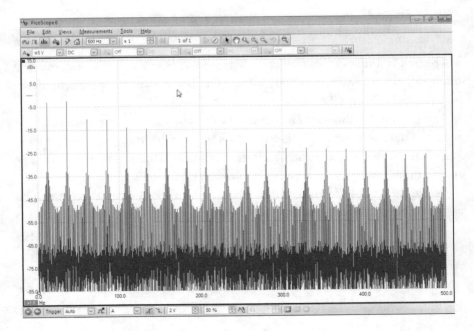

Figure 2-13   *Pseudo-random noise spectrum.*

Summary

The chapter began with a review of the key hardware features on the Pyboard. I then provided some background on how microcontroller programs are constructed, which was a prelude to a detailed discussion on interrupts. I stressed how interrupts are very important for well-designed microcontroller applications as would be run on a Pyboard.

Some Pyboard LED demonstrations followed using the REPL prompt for command entry. I also showed you how to use a program (Python script) to control two of the Pyboard LEDs.

A section followed that detailed how similar Python and MicroPython routines could function in a program, with the MicroPython routines being smaller and more efficient. A hardware interrupt demonstration followed, which showed how the Pyboard USR push button could be used to force an interrupt.

I next introduced some basic concepts on how to control Pyboard timers with REPL commands. These commands were further expanded into a full program to generate interrupts, which blinked two of the Pyboard LEDs.

The chapter concluded with a detailed discussion on how to use both the analog-to-digital converter (ADC) and the digital-to-analog converter (DAC).

3

Interfaces, Files, and Libraries

In this chapter I will continue to explore some more Pyboard hardware features, including pulse width modulation (PWM) and three bit-serial communication interfaces that are key to implementing communication links with external devices. I will also show you how to load modules into the Pyboard and explore how the boot process works. Note that I will use the term *module* to generically represent any digital content such as a function definition, file, or library.

Pulse Width Modulation

PWM is a method of controlling external devices such as servos or direct current (DC) motors. It is essentially a process where the on time or high state of a digital signal is directly proportional to some desired condition for an external device. The extent and repetition rate vary, depending on the type of device being controlled. A standard analog servo typically requires a 50-Hz digital pulse whose width varies between 1.0 and 2.0 ms as shown in Figure 3-1.

A PWM analog DC motor control signal would vary considerably from the waveform shown in the figure. The DC motor control would typically have its on-time pulse width vary from 0 to 100 percent of the waveform period, representing 0 to 100 percent equivalent change in motor speed. The servo PWM range of 1 to 2 ms typically represents a –90 degree to 90 degree position change for a standard analog servo control arm. This same range would represent a maximum speed counterclockwise (CCW) to maximum speed clockwise (CW) range, if a continuous rotation (CR) servo was being controlled.

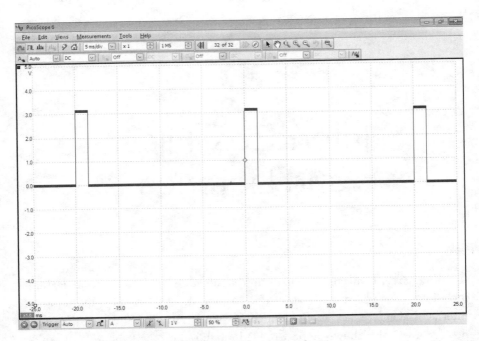

Figure 3-1 *Standard analog PWM control signal waveform.*

The Pyboard's blue light-emitting diode (LED) can be controlled using a PWM signal, which means its intensity can be precisely controlled. This is not true of the other onboard LEDs whose state can only be on or off. The following program demonstrates how the blue LED intensity can be controlled:

```
# no magic numbers
blueLED = 4
led = pyb.LED(blueLED)

intensity = 0

while True:
        # slowly increase LED intensity with remaindering operator
    intensity = (intensity + 1) % 255
    led.intensity(intensity)
    pyb.delay(30)
```

I added this code to the main.py file and rebooted the Pyboard. After the restart, I observed the blue LED repeatedly slowly cycle from off to maximum brightness, thus proving the PWM functioned as expected.

I will later use the PWM function in the robotic car project to control two CR servos that provide the moving force for the car.

Bit-Serial Ports

The Pyboard supports three bit-serial protocols that allow for serial communication between it and companion devices or other computing systems.

UART Serial Protocol

The first bit-serial protocol to be discussed is the standard Universal Asynchronous Receive Transmit (UART) that uses two pins in the Pyboard (not counting ground) and is shown in the block diagram in Figure 3-2.

This protocol needs no clock signal, as is indicated by the word *asynchronous* in the name. The Pyboard transmits data on the pin named TX and receives on RX. Remarkably, there are five UART buses physically implemented on the Pyboard. Table 3-1 shows the Pyboard's pin designations for all seven of the UART (CAN) TX and RX buses.

NOTE *Bus 5 was not listed as you might have expected. One of two controller area network (CAN) ports is dedicated to that bus number. I will discuss CAN ports later in this book.*

Bus Number	Description	Pyboard Pin Number
1	TX, RX	X9, X10
2	TX, RX	X3, X4
3	TX, RX	Y9, Y10
4	TX, RX	X1, X2
6	TX, RX	Y1, Y2

Table 3-1 *UART Signal Lines*

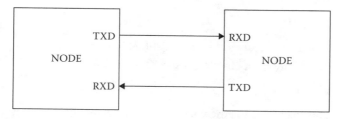

Figure 3-2 *UART block diagram.*

The UART protocol is used primarily for data communication and needs no master or slave, as would be needed in a control protocol such as I2C or SPI, which are discussed in later sections.

The pyb library contains a comprehensive UART class definition that fully supports this serial protocol. The following code snippet demonstrates how UART objects may be instantiated and initialized. The code listing has verbose comments to help you understand the statements.

```
from pyb import UART
# instantiate a uart object with a 9600 baudrate
uart = UART(1, 9600)
# initialize the uart object with the number of data, stop and parity bits
# data bits maybe 7 or 8; stop 1 or 2; and parity None or 0 (even), 1 (odd).
uart.init(9600, bits=8, stop=1, parity=None)
```

Many other methods are available to assist with creating practical UART code. Some of these are

```
uart.any()
```

Returns the number of characters waiting in the receive buffer.

```
uart.recv(recv, *, timeout=5000)
```

Receives data on the bus for which the uart object was originally instantiated. recv can be an integer, referring to the number of bytes to be received, or as buffer reference to be filled with received bytes. timeout is the timeout in milliseconds to wait for any incoming data.

Return value: If recv is an integer, then the returned value is a reference to a new buffer of received bytes; otherwise, it is the same buffer that was passed in to recv.

```
uart.send(send, *, timeout=5000)
```

Sends data on the bus for which the uart object was originally instantiated. send is the data to send (an integer to send or a buffer object). timeout is the timeout in milliseconds to wait for the send.

Return value: None.

```
uart.readinto(buf)
```

Stores received characters into the predefined buffer, buf.

Return value: Number of characters stored in the buffer.

```
uart.deinit()
```

Turns off the bus for which the uart object was originally instantiated.

Practical UART Demonstration

I believe it is important to demonstrate how the UART protocol functions in real life. After some consideration, I decided to use a global positioning system (GPS) module that continuously generates data and outputs it as a stream via the UART protocol. Figure 3-3 shows the Ultimate GPS module version 3 connected to the Pyboard using a small, solderless breadboard. The Ultimate GPS breakout board is part number 746 and is available at www.adafruit.com.

I first tested the GPS module to confirm that it was functioning well. This is a very critical step to do before trying to use it with the Pyboard. If it was not

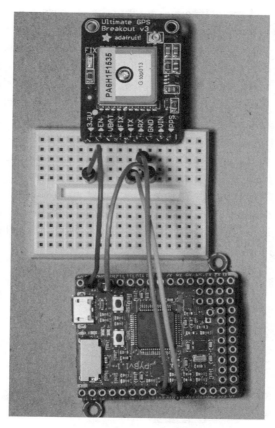

Figure 3-3   *Ultimate GPS breakout board connected to the Pyboard.*

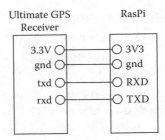

Figure 3-4 *Schematic for the interconnection between the RasPi and Ultimate GPS module.*

working as expected and you connected it to the Pyboard, you would then be in the predicament of determining if the Pyboard interface was okay, whether or not the software was functioning as designed, or if the GPS module itself was not working. Eliminating one or several potential trouble spots is always a good idea. I conducted the initial test using a Raspberry Pi (RasPi) on which I installed and ran the CuteCom terminal program. The GPS module was connected to the RasPi as shown in the schematic in Figure 3-4.

It only requires four leads to set up the connection. The first lead connects the RasPi's 3V3 pin to the GPS module's 3.3-V pin. The second lead goes from the RasPi TXD pin to the GPS module RX pin. The third lead goes from the RasPi RXD pin to the GPS module's TX pin. The fourth and final lead simply connects the RasPi ground or common to the ground on the GPS module. The schematic for interconnecting the Pyboard is almost identical to the RasPi interconnection schematic and is shown in Figure 3-5.

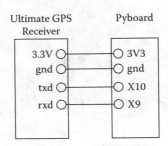

Figure 3-5 *Schematic for the interconnection between the Pyboard and Ultimate GPS module.*

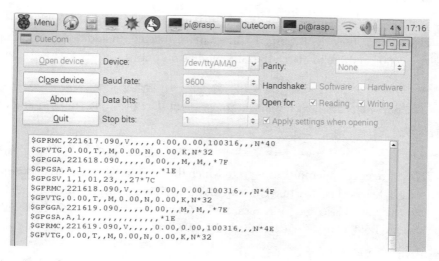

Figure 3-6 *CuteCom screen capture of GPS data stream.*

A snapshot of the output data stream between the RasPi and the GPS module can be seen in Figure 3-6, which was captured while the CuteCom program was running.

This figure shows that the GPS was working properly and it was generating a variety of GPS "sentences" that make up the data stream. In this case, the chip antenna that is internal to the GPS module was sufficiently sensitive to receive the GPS satellite signals without the need for an external antenna. I provide a much more detailed discussion on the GPS module and related accessories in Chapter 8 that deals with a GPS project.

The next step is create MicroPython test software that will instantiate a UART object and allow for the presentation of sample GPS data. The following commands were issued at the read-eval-print-loop (REPL) prompt, which I found to be quite adequate to test the UART protocol. Just connect the GPS module to the Pyboard using the schematic in Figure 3-4 as a guide and enter the following commands at the REPL prompt. Note that I have provided some further explanatory details about each command right before the actual command.

Importing the pyb library containing the UART class must be the first command:

```
import pyb
```

A UART object named `uart` is then instantiated. It is associated with bus #1 because that is how the RX and TX lines are connected. In addition, it is set up with a 9600 baudrate to match the GPS module output.

```
uart = pyb.UART(1, 9600)
```

The uart object is next initialized to use eight data bits, one stop bit, and no parity bit.

```
uart.init(9600, bits=8, stop=1, parity=None)
```

Next, I needed a memory buffer into which the GPS data stream can be stored. I used a bytearray because that container is best suited for this requirement and it is easy to create. I also determined through a bit of experimentation that it should be sized to hold 63 bytes to best match the GPS data stream.

```
myFrame = bytearray(63)
```

The data can easily be stored in the bytearray using this next command. It is perfectly possible and quite likely to receive "broken" GPS sentences because they are not particularly synchronized to this method of data storage. Just repeat the command until you receive a complete and "unbroken" GPS sentence. After being run, the command should display the number 63, which is the number of bytes received and stored.

```
uart.readinto(myFrame)
```

Next, enter the bytearray name, myFrame, to display the received GPS data:

```
myFrame
```

This next line shows a sample of the myFrame data:

```
bytearray(b'$GPGGA,230053.000,2656.6526,N,08213.2058,W,1,07,1.27,-10.5,M,-2')
```

I will defer explaining how to decipher this GPS "sentence" until I discuss the GPS project, but I will say that the GPS coordinates shown here are accurate and represent the precise location and elevation at which the GPS module was located when the reading was captured.

To reiterate, this demonstration was not about how to receive GPS signals, but how to use a UART protocol to transfer data between a Pyboard and an UART-compatible external device. This finishes the first bit-serial protocol discussion, and the next will be about the I2C protocol.

I2C Serial Protocol

The second bit-serial protocol is the inter-integrated circuit interface, or I2C (pronounced "eye-two-cee" or "eye-squared-cee"), which is also known as a synchronous serial data link. A clock signal is needed because it is synchronous, unlike the previously discussed UART protocol, which is inherently self-synchronizing. Figure 3-7 is a block diagram of the I2C interface showing one master and one slave. This configuration is known as a multidrop or bus network.

I2C supports more than one master as well as multiple slaves. This protocol was created by the Philips Company in 1982 and is a very mature technology, meaning it is extremely reliable. Only two lines are used: SCLK for serial clock and SDA for serial data.

The Pyboard implements two separate I2C buses: 1 and 2. Table 3-2 shows the Pyboard's pin designations for the clock and data lines for both buses.

Signal Name	Description	Pyboard Pin Number
SCL	Bus 1 Clock	X9
SDA	Bus 1 Data	X10
SCL	Bus 2 Clock	Y9
SDA	Bus 2 Data	Y10

Table 3-2 *I2C Signal Lines*

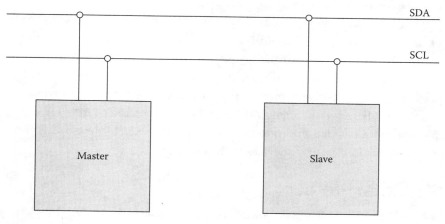

Figure 3-7 *I2C block diagram.*

The pyb library contains a complete I2C class definition that fully supports this serial protocol. The following code snippet demonstrates how both master and slave I2C objects may be instantiated and initialized. The code listing has verbose comments to help you understand the statements:

```
from pyb import I2C
# create a master i2c object on bus 1
i2c = I2C(1, I2C.MASTER)
#initialize the i2c object with a 50 KHz  data rate
i2c.init(I2C.MASTER, baudrate=50000)
# create a slave i2c object on bus 1
i2c_slave = I2C(1, I2C.SLAVE)
# initialize the i2c_slave object with a hex address of 0x42
i2c_slave.init(I2C.SLAVE, addr=0x42)
```

The slave address is set by the actual hardware configuration. Most I2C slave devices have a provision to change their addresses, which helps prevent address collisions when multiple slave devices are simultaneously connected on the parallel bus.

Sending and receiving ASCII data between a master and slave is quite simple, as this following code snippet reveals:

```
# send 5 bytes to a slave at address 0x42
i2c.send('hello', 0x42)
# receive 5 bytes
data = i2c.recv(5)
```

And in case you do not know the slave addresses, you can always direct the master to reveal them by using i2c.scan().

It is also useful to check if a slave device is ready to communicate with the master. The command i2c.is_ready(0x42) will return a Boolean indicating whether or not the I2C slave at address 0x42 is ready.

Many other methods are included in the pyb I2C class that could be useful in developing programs that incorporate I2C devices with a Pyboard. I would refer you to the MicroPython application programming interface (API) for the I2C class for more detailed information. I also used the API in developing project programs that interface I2C devices to the Pyboard.

SPI Serial Protocol

The third and final bit-serial protocol that I will discuss is the Serial Peripheral Interface (SPI) that is shown in the block diagram in Figure 3-8.

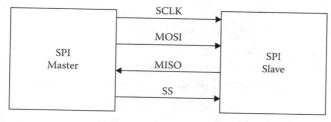

Figure 3-8 *SPI block diagram.*

The SPI interface (pronounced "spy" or "ess-pee-eye") is also known as a synchronous serial data link. It is also a full duplex protocol, meaning data can be simultaneously sent and received between the host and slave. SPI is also referred to as a synchronous serial interface (SSI) or a four-wire serial bus.

The Pyboard implements two separate SPI buses: 1 and 2. The eight interconnecting SPI signal lines are detailed in Table 3-3.

As was the case with the I2C protocol, the pyb library contains a complete SPI class definition that fully supports this serial protocol. The following code snippet demonstrates how a master SPI object may be instantiated and initialized.

```
from pyb import SPI
spi = SPI(1, SPI.MASTER, baudrate=600000, polarity=1, phase=1, crc=0x7)
```

The only required arguments in the statement instantiating the SPI object are the bus number and the constant specifying if the object is a master or a slave.

Signal Name	Description	Pyboard Pin Number
MOSI	Bus 1 Master Out Slave In	X8
MISO	Bus 1 Master In Slave Out	X7
SCK	Bus 1 Clock	X6
/SS	Bus 1 Slave Select	X5
MOSI	Bus 2 Master Out Slave In	Y8
MISO	Bus 2 Master In Slave Out	Y7
SCK	Bus 2 Clock	Y6
/SS	Bus 2 Slave Select	Y5

Table 3-3 *SPI Signal Lines*

In this example, the SPI object is named `spi` and it is set to operate from bus 1 as a master. The baudrate is optionally set at 600 kHz with polarity set at 1, which means its idle state when no data is being transferred is high. A 0 polarity would consequently mean the idle state is low. The phase set at 1 means that one edge or signal transition is used for detecting signal line changes. Phase set at 2 would mean two transitions would be used to detect signal line transitions. Finally, the crc set at 0x7 means that a seventh-order polynomial is to be used to detect and correct any transmission line errors. The crc may be set to None, which means no error detection or error correction would be implemented.

Here are some additional methods that are used to send and receive data over a SPI link:

```
# send 4 bytes and receive 4 bytes
data = spi.send_recv(b'1234')
# create a container named buf to receive some data
buf = bytearray(4)
# send 4 bytes and receive 4 into buf
spi.send_recv(b'1234', buf)
# send/recv 4 bytes from/to buf
spi.send_recv(buf, buf)
```

As was the case for the I2C class, many other methods are included in the pyb SPI class that could be useful in developing programs that interface SPI devices with a Pyboard. I would refer you to the MicroPython API for the SPI class for more detailed information.

All three serial protocols described in these sections are implemented in hardware, meaning there is actual silicon dedicated to performing the protocol functions. This is the most efficient and fastest way to provide these serial interfaces, but it is not the only way. These same serial interfaces can be implemented using uncommitted general-purpose input/output (GPIO) pins and software. This would provide nearly the same functionality but would not be as fast compared with the hardware implementation. The term "bit-banging" is often used to describe this approach. Sometimes you must use bit-banging when the hardware is not available. Fortunately for all the projects described in this book, bit-banging will not be required because the Pyboard has more-than-adequate hardware implementations available.

It is now time to change the discussion from hardware-oriented features to software-focused features.

Directory Structures

Upon a normal boot-up, MicroPython initializes a root directory in its flash ROM named /flash, as can be seen in Figure 3-9.

To create this list, I first needed to import the sys module and then entered sys.path to have MicroPython enumerate the initialized directories. You can clearly see the root directory, /flash, listed. In the figure you can see that I also imported the os module and subsequently entered os.listdir('/flash') to get a listing of all the files contained within the /flash directory. You will not normally need use the listdir function to display the content of the root directory, as it is automatically replicated by the USB drive on the host computer, which is PYBFLASH for my situation using a MacBook Pro to connect to the Pyboard. The important fact to realize is that the /flash directory on the Pyboard and the USB drive PYBFLASH on the host are one and the same. It is just that /flash directory can only be accessed through MicroPython, whereas the PYBFLASH folder is accessed and manipulated using the standard host utilities. I found it much easier to simply use the PYBFLASH folder for any file/directory entries or edits. You will also find out that this duality of directory/folder access holds true when the micro SD card is used on the Pyboard.

In Figure 3-10, I show the contents of the PYBFLASH folder as displayed on the Mac host.

The contents of the /flash directory and the PYBFLASH folder are identical, as may be seen by comparing the two figures, which further proves the duality of the directories.

```
donnorris — screen /dev/tty.usbmodem1412 ▶ SCREEN — 90×24

MicroPython v1.5.1-66-g66b9682 on 2015-12-04; PYBv1.1 with STM32F405RG
Type "help()" for more information.
>>> import sys
>>> sys.path
['', '/flash', '/flash/lib']
>>> import os
>>> os.listdir('/flash')
['main.py', 'pybcdc.inf', 'README.txt', 'boot.py', '._.Trashes', '.Trashes', '.fseventsd',
 '.TemporaryItems', '._.TemporaryItems', '._main.py', 'extDir', 'extLib.py']
>>>
```

Figure 3-9 *sys path listing.*

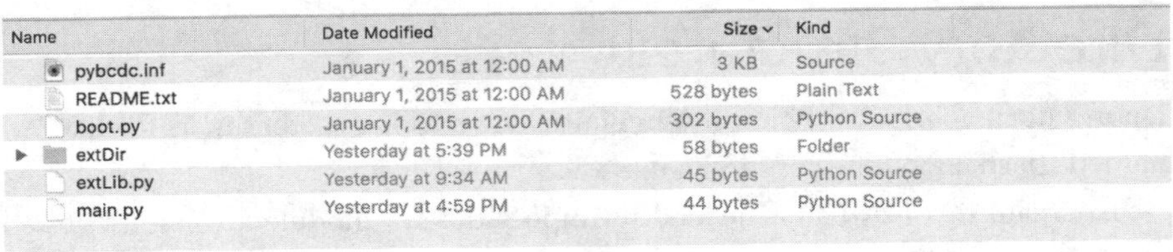

Name	Date Modified	Size ⌄	Kind
⦿ pybcdc.Inf	January 1, 2015 at 12:00 AM	3 KB	Source
README.txt	January 1, 2015 at 12:00 AM	528 bytes	Plain Text
boot.py	January 1, 2015 at 12:00 AM	302 bytes	Python Source
▶ ▦ extDir	Yesterday at 5:39 PM	58 bytes	Folder
extLib.py	Yesterday at 9:34 AM	45 bytes	Python Source
main.py	Yesterday at 4:59 PM	44 bytes	Python Source

Figure 3-10 *PYBFLASH contents.*

Importing a Module

I will start by saying that because the Pyboard lacks an Internet connection, this places a severe constraint on how to import modules into the board's memory. I will discuss several approaches, with the easiest one just placing the module into a file and storing that file in the USB drive, which is automatically created when you plug the Pyboard into your host computer. The next section takes you through a step-by-step procedure on how to accomplish this approach.

Using a File for an Import

I will simply use a small function definition for this example, as it will adequately demonstrate how the file import works. The following is a function definition stored in a file named extLib.py, which I created using the nano text editor on my MacBook Pro. You can use any text editor you have—just ensure that no special characters or unique formatting is inadvertently added.

```
def hello():
    print("Hello from extLib")
```

This file must be added to the USB drive that is created on the host when the USB cable from the Pyboard is plugged in. This drive is named PYBFLASH on my Mac, and it should have the same or similar name for Windows and Linux hosts. You next have to change the contents of the main.py file, as was done in the previous chapter. Add the following to main.py:

```
import extLib
extLib.hello()
```

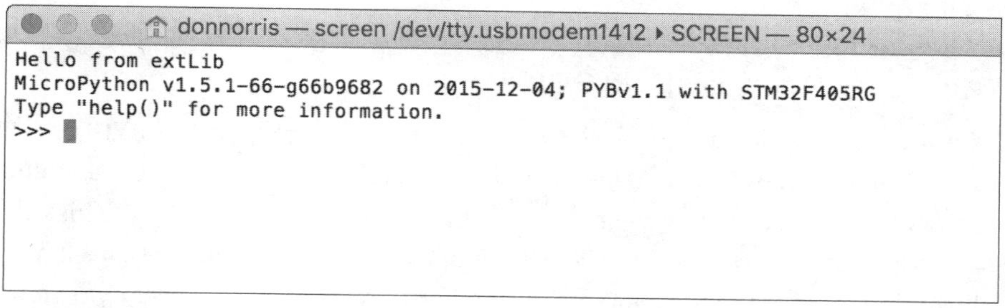

```
● ● ●    ⌂ donnorris — screen /dev/tty.usbmodem1412 ▸ SCREEN — 80×24
Hello from extLib
MicroPython v1.5.1-66-g66b9682 on 2015-12-04; PYBv1.1 with STM32F405RG
Type "help()" for more information.
>>> █
```

Figure 3-11   *Import example.*

Next, eject the USB drive from the host and unplug the USB cable connecting the Pyboard. You should now plug the cable back into your host and reconnect to the Pyboard using a terminal program. Figure 3-11 is a screen shot of this reestablished connection, demonstrating that the extLib file was properly imported, and the hello function defined in it was also automatically executed due to it being added to the main.py file in the PYBFLASH drive.

I next deleted the contents of the main.py file on the PYBFLASH drive and rebooted the Pyboard in order to demonstrate how to both manually import the extLib file and execute the hello() all using the REPL mode. Figure 3-12 shows the two commands I entered and the resulting output, which was precisely the same as shown in the previous figure.

The next approach discusses how to import a module from a subdirectory to the PYBFLASH root directory.

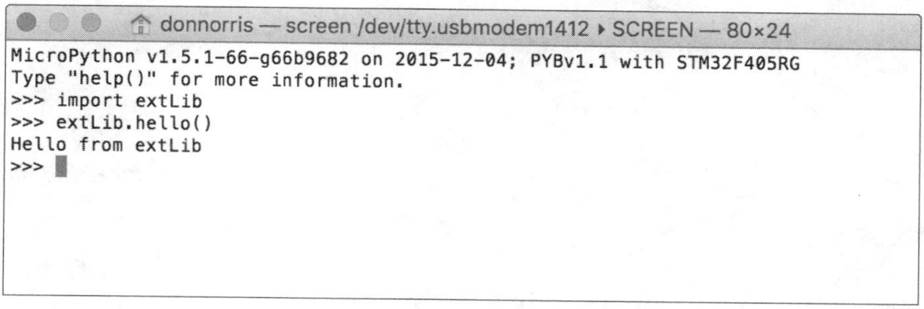

```
● ● ●    ⌂ donnorris — screen /dev/tty.usbmodem1412 ▸ SCREEN — 80×24
MicroPython v1.5.1-66-g66b9682 on 2015-12-04; PYBv1.1 with STM32F405RG
Type "help()" for more information.
>>> import extLib
>>> extLib.hello()
Hello from extLib
>>> █
```

Figure 3-12   *Manual import and function execution.*

Importing a Module from a PYBFLASH Subdirectory

The first step in this approach is to create a new directory in the PYBFLASH drive. I added a new folder named extDir to the folder using the Mac File menu option. I next added a new file named __init__.py to this subdirectory. This odd name reflects how MicroPython—or Python for that matter—instantiates a new object. In this case the file contents are similar to what was done in the previous example.

```
def hello():
    print('Hello from the extDir directory')
```

This particular code may be run from the REPL prompt by entering these commands:

```
import extDir
extDir.hello()
```

Figure 3-13 shows the result of entering the two commands.

Now, I bet you are thinking: What is the advantage of using the external directory approach to importing a module? One neat advantage is you can easily add function definitions to the module without making any changes to the existing function definitions, while still allowing the grouping of similar-type functions. Thus, it is entirely possible to have certain custom math functions together in a single subdirectory and have custom list-processing functions in another directory. I have made

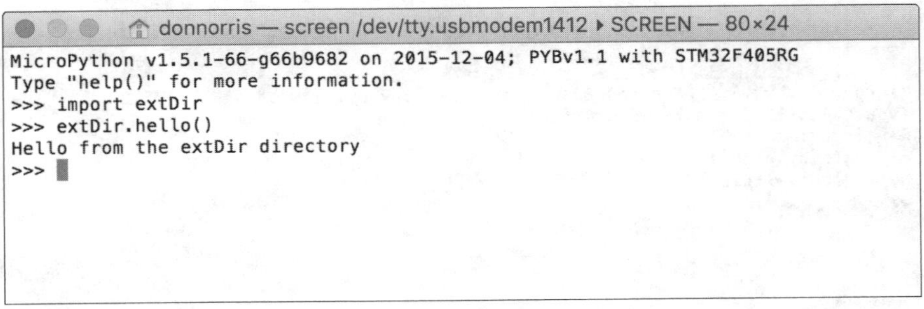

```
MicroPython v1.5.1-66-g66b9682 on 2015-12-04; PYBv1.1 with STM32F405RG
Type "help()" for more information.
>>> import extDir
>>> extDir.hello()
Hello from the extDir directory
>>>
```

Figure 3-13 *Import module from a subdirectory.*

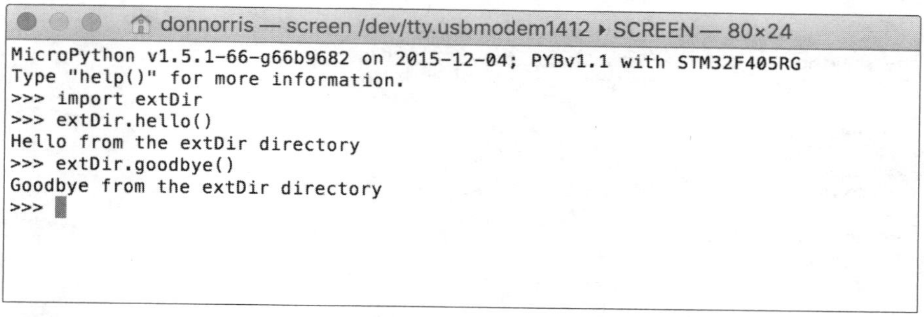

```
 ●  ●  ●      donnorris — screen /dev/tty.usbmodem1412 ▸ SCREEN — 80×24
MicroPython v1.5.1-66-g66b9682 on 2015-12-04; PYBv1.1 with STM32F405RG
Type "help()" for more information.
>>> import extDir
>>> extDir.hello()
Hello from the extDir directory
>>> extDir.goodbye()
Goodbye from the extDir directory
>>> ▊
```

Figure 3-14   *Edited external directory extDir import example.*

a simple demonstration in the following code of how easy it is to add on to the __init__.py module in the extDir:

```
def hello():
    print('Hello from the extDir directory')

def goodbye():
    print('Goodbye from the extDir directory')
```

Figure 3-14 shows the results after I reran the import and function commands. The next approach on how to import modules relies on the use of an SD card.

Using the SD Card for an Import

The Pyboard has a socket mounted on it where you can plug in an optional micro SD card. I inserted an 8-GB card that I named MP_SD and created a subdirectory named extSD. An interesting action happens when you connect a Pyboard with a micro SD card inserted in its socket. A folder reflecting the SD card root directory will appear on the host computer, which in this case was named MP_SD because I had previously set the SD card as its volume name. The normal PYBFLASH folder is not created when an SD card is inserted in the Pyboard's micro SD socket. The MP_SD folder on the host and the /sd directory on the Pyboard are the entities reflecting the duality of the file system as previously discussed. I also confirmed that the new root directory on the Pyboard was /sd by fetching the current working directory (CWD) via the os.getcwd() command at the REPL prompt. This is shown in Figure 3-15, along with an extSD import statement and two function calls that again demonstrate how functions can be called from an imported external directory.

The resulting print statements are almost identical to what was shown in Figure 3-6, except I changed the name of the imported subdirectory.

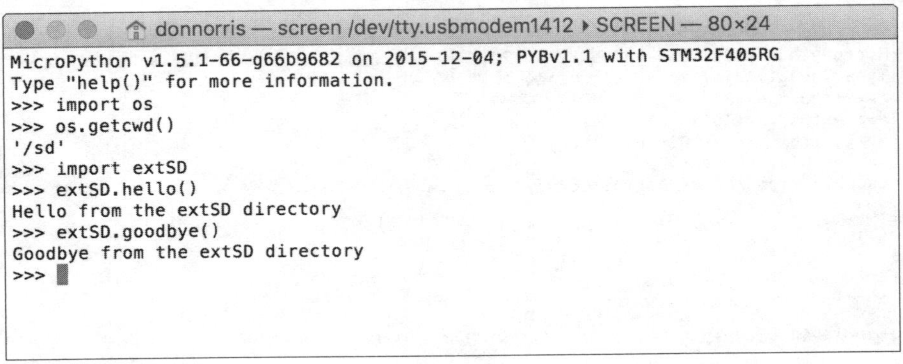

```
● ● ●      ⚙ donnorris — screen /dev/tty.usbmodem1412 ▸ SCREEN — 80×24
MicroPython v1.5.1-66-g66b9682 on 2015-12-04; PYBv1.1 with STM32F405RG
Type "help()" for more information.
>>> import os
>>> os.getcwd()
'/sd'
>>> import extSD
>>> extSD.hello()
Hello from the extSD directory
>>> extSD.goodbye()
Goodbye from the extSD directory
>>> █
```

Figure 3-15 *SD card import and function calls.*

This section concludes my brief introduction to importing modules using MicroPython. I will turn to the boot process next, as this, too, can have a direct effect on how modules are executed on the Pyboard.

Boot Process

There are several ways that the Pyboard can be booted or started up from a powered-off state:

- standard—boot.py file executed first USB configured main.py file executed
- safe—does not execute any scripts (files) at boot-up
- filesystem reset—restores the filesystem to the original factory state boots into safe mode

A Pyboard normally powers up via the standard boot mode. To boot into one of the two nonstandard modes, you must use the following procedure:

1. Press and hold the USR push button when powering up the Pyboard.

2. Press and release the RST push button.

3. The onboard green and orange LEDs will continually light in a binary sequence as follows:

   a. Green LED only—standard boot

   b. Orange LED—safe boot

   c. Green and orange LEDs—filesystem reset

4. Release the USR button when the desired boot mode is displayed by the LEDs.

It is also important to discuss the contents of the boot.py script, as that has a significant effect on how the user can interact with the Pyboard. The following is a listing of the boot.py file:

```
# boot.py -- run on boot-up

# can run arbitrary Python, but best to keep it minimal

import machine

import pyb

#pyb.main('main.py') # main script to run after this one

#pyb.usb_mode('CDC+MSC') # act as a serial and a storage device

#pyb.usb_mode('CDC+HID') # act as a serial device and a mouse
```

The file is quite simple, as only the machine and pyb libraries are imported. The remaining statements are commented out. In a standard boot process, any content in the main.py script will always be run after the boot.py script. The next two commented statements are exclusive; there can only be one or the other, but not both. The statement pyb.usb_mode('CDC+MSC') is also the standard boot default and does not need to be explicitly uncommented. This mode creates a USB connection with a host computer and sets up a mass storage device, that is, the USB flash drive named PYBFLASH for my case. The next statement, if uncommented, would also establish a USB connection but no USB flash drive. Instead, the Pyboard would act as a human interface device (HID), which in this situation is either a mouse or keyboard. This last mode comes in handy when attempting to load the Pyboard's micro SD card during the boot process in lieu of the regular USB flash drive. I will further discuss this boot mode when I cover how to set up a data logger project with the SD card as the data-recording device.

Error Reporting by the LEDs

The Pyboard will notify you of two error conditions by blinking the onboard LEDs as follows:

- When the red and green LEDs alternately flash, this normally indicates an error while attempting to execute the code stored in main.py. It is recommended that the code be debugged using the REPL prompt.

- A hard fault condition is present when all four LEDs slowly blink. You must do a hard reset by pressing the RST push button to recover from this condition. It is recommended that you first close the serial link connection program and eject the USB drive before doing the hard reset. Note that it may even require a filesystem reset boot if the hard reset also fails.

There may be additional LED error reporting provided in future firmware revisions, but for now that is all that is currently available.

Libraries

I will finish this chapter with a brief discussion of the libraries that are either built into MicroPython or can be loaded if needed.

Standard Libraries

The following are the standard Python libraries built into MicroPython. I am assuming that the word *standard* refers to libraries that conform to the Python 3 API and can be used in MicroPython in exactly the same way as they would be used in Python 3.

- cmath—mathematical functions for complex numbers
- gc—control the garbage collector
- math—mathematical functions
- os—basic "operating system" services
- select—wait for events on a set of streams
- struct—pack and unpack primitive data types
- sys—system-specific functions
- time—time-related functions

You may also realize I have used some of the methods from these libraries in code already presented in the book.

Customized Python Libraries

MicroPython also has a set of Python libraries that have been customized in some fashion to work efficiently with the language. The library names have been prepended with the letter *u* to distinguish them from the regular Python 3 variety.

A special import process kicks in if you use the regular name, such as import json instead of import ujson. The import json statement will first search for a file named json.py and, not finding that, will continue to search for a directory named json. In case neither one of those is found, the built-in ujson library will be automatically loaded.

- ubinascii—binary/ASCII conversions
- uctypes—access C structures
- uhashlib—hashing algorithm
- uheapq—heap queue algorithm
- ujson—JSON encoding and decoding
- ure—regular expressions
- usocket—socket module
- uzlib—zlib decompression

Pyboard-Specific Library and Classes

There is one library that is built into MicroPython but is only specific to the language when it is run on the Pyboard. You already have encountered the pyb library in code that I have presented. This library contains a lot of functionality, thus making the Pyboard a useful microcontroller development platform. These functionalities include but are not limited to:

- Time-related functions
- Reset-related functions
- Interrupt-related functions
- Power-related functions
- Miscellaneous functions

The specific classes contained in the pyb library include but are not limited to the following:

- class Accel—accelerometer control
- class ADC—analog-to-digital conversion
- class CAN—controller area network communication bus
- class DAC—digital-to-analog conversion

- class ExtInt—configure I/O pins to interrupt on external events
- class I2C—a two-wire serial protocol
- class LCD—LCD control for the LCD touch-sensor pyskin
- class LED—LED object
- class Pin—control I/O pins
- class PinAF—pin alternate functions
- class RTC—real-time clock
- class Servo—three-wire hobby servo driver
- class SPI—a master-driven serial protocol
- class Switch—switch object
- class Timer—control internal timers
- class TimerChannel—set up a channel for a timer
- class UART—duplex serial communication bus
- class USB_VCP—USB virtual comm port

I have already included some of these classes in my demonstration code. Much more will follow in the project discussions. Class development is a dynamic process, with new ones constantly being developed and tested. Some recent classes include:

- Network—for network configuration
- CC3K—SPI interface for the CC3K wireless LAN module
- WIZNET5K—Ethernet adapters based on the W5200 and W5500 chipsets

I would urge you to occasionally check the www.micropython.org website to see what new classes have been added and what new devices are supported by the MicroPython language.

MicroPython Library

There is also a specialized library, aptly named micropython, that will allow access and control to certain functions within the actual MicroPython language. As of this writing, only three methods seem to be available.

NOTE   *This information is reprinted directly from the www.micropython.org website. I did not want to misinterpret this information, as any time you delve into a language's internals you have a good chance of causing some issues.*

```
micropython.mem_info([verbose])
```

Print information about currently used memory. If the verbose argument is given, then extra information is printed.

The information that is printed is implementation dependent, but currently includes the amount of stack and heap used. In verbose mode it prints out the entire heap, indicating which blocks are used and which are free.

```
micropython.qstr_info([verbose])
```

Print information about currently interned strings. If the verbose argument is given, then extra information is printed.

The information that is printed is implementation dependent, but currently includes the number of interned strings and the amount of RAM they use. In verbose mode it prints out the names of all RAM-interned strings.

```
micropython.alloc_emergency_exception_buf(size)
```

Allocate size bytes of RAM for the emergency exception buffer (a good size is around 100 bytes). The buffer is used to create exceptions in cases where normal RAM allocation would fail (e.g., within an interrupt handler) and therefore gives useful traceback information in these situations.

A good way to use this function is to put it at the start of your main script (e.g., boot.py or main.py), and then the emergency exception buffer will be active for all the code following it.

This last section concludes this chapter. The next chapter starts the actual book projects, which hopefully will reinforce all the preliminary information provided in these first three chapters. Feel free to review these chapters as you progress through the remaining ones. I have a strong feeling that you will need to do such a review.

Summary

A pulse width modulation (PWM) discussion started the chapter. PWM is important for power control, and I provided a simple LED intensity demonstration to illustrate the concept.

Three sections followed detailing three bit-serial communications protocols that are implemented on the Pyboard and supported by MicroPython classes. These protocols are UART, I2C, and SPI, and all are reasonably explained in the chapter.

I next shifted gears a bit and discussed the directory structure created when you plug in the Pyboard's USB cable into a host computer. Included in this discussion were several ways to import files/programs into the flash memory. The boot process was also covered, including an explanation on how to interpret the onboard LED signals to detect errors.

The chapter concluded with a comprehensive discussion regarding the various libraries that are available to assist with your MicroPython program development.

4

Let Ball Detector

In this chapter I will show you how I designed and built an interesting project that functions as a let ball detector. A let ball is a tennis term that refers to when a tennis ball on a serve just strikes the top of the net separating tennis players yet continues on a straight-line trajectory. The ball must also land into the correct service box diagonal to the server. Figure 4-1 is a photograph of a let ball strike.

Such a serve is a "do over" and does not count as a fault against the player causing the let ball. Sometimes it is quite difficult to detect when a let ball situation happens, which is the reason why there is a referee stationed at one end of the net in professional tennis tournaments to detect such occurrences. My project replaces the referee with Pyboard, which will flash a light-emitting diode (LED) array if it detects a let ball strike. Of course, my project cannot actually replace the highly skilled senses of a professional tennis referee, but it does a reasonable job that will suit most amateur players, and you don't have to pay it anything.

I do want to explain what a net ball is, as it sometimes can be confused with a let ball. A net ball also strikes the net, but fails to land in the correct service box. The server does not get an opportunity to replay the serve and is assessed a fault. A second net ball on the same service will result in a double fault.

Figure 4-1 *Let ball strike.*

79

This project will take advantage of the Pyboard's built-in accelerometer to sense the vibrations that occur in the net as the tennis ball grazes the top of the reinforced canvas strip that supports it. I will begin this project with a short presentation of my approach to designing a project, as that will provide at least one framework to guide you in a project design. This approach may not be the best one, but it has served me well in the successful completion of many embedded projects.

Initial Design

I always start a project by writing down the requirements that must be met upon project completion. This step is especially vital if you are under contract to a client to design and build a specific project. Putting the requirement in writing ensures that certain clarity is provided such that all parties read and understand them and ultimately agree to them. This statement of requirements is dynamic because the project steps and requirements can and do change as facts are discovered and realities are firmed up as to what can actually be achieved in the allotted timeframe. I strongly recommend that you do not skip this beginning step even if you are doing this project on your own without any other committed parties involved. Explicitly stating project requirements ensures you have a good understanding of what must be accomplished. Furthermore, if you fail to meet all the requirements, you have the option of recording why the requirement was not met and what you might do in the future to overcome that failure.

Project Requirements

Project requirements must start with the overall project description, which in this case is to detect all let ball occurrences during a tennis match. Of course, two other key requirements are that a Pyboard is to be used and MicroPython will be the development language. Additional requirements usually result from considering the environment in which the device(s) will be used. For this case, I am pretty sure that the Pyboard will have to be portable, which means battery operation. There should also be some type of visual indication to the players that a let ball has happened and perhaps an optional audio indicator. The latter should have an on/off switch, as tennis players are usually very courteous and considerate of other players on adjoining courts. The device should also be able to be quickly and securely attached to the net in a temporary manner without altering or damaging it. The following is a summary list of the initial requirements:

- Detect tennis let ball occurrences
- Pyboard microcontroller
- MicroPython development language
- Battery operated
- Visual and optional audio alert
- Secure, nondamaging attachment to net

In reality, other related project requirements usually are included, including not but limited to, total development time, material costs, regulatory issues, and available resources including manning and support equipment.

A quick review of the six listed requirements shows that all should be readily achievable given that the actual material is on hand. I did a quick inventory of my available parts and realized that I did not have the following items:

- Audio alert device
- Pyboard case
- Battery pack
- Attachment device

At that point I quickly ordered the missing pieces so as to minimize any delay in project completion. Reviewing your project material needs versus your on-hand resources will help you identify any shortfalls and allows time for you to order the needed parts. This is especially important if you happen to require long-lead material, which could take many weeks to obtain. I was not too concerned about the attachment device, as that would only require readily available material that I could purchase from a local home improvement store.

The next step in this design process is to create a prototype, which can be simple sketches or drawings or even some preliminary hardware that can be "breadboarded."

Prototype

A prototype is simply a starting point for the design, which will fulfill the project requirements. I like to use a sketch as my preferred choice as it quite easy to

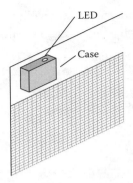

Figure 4-2 *Initial prototype sketch.*

modify and alter as necessary to incorporate all the desired features. Figure 4-2 is a sketch of the initial design.

The sketch is not really detailed or very complete, as it is only a starting point and will be changed as you continue with the design process. This initial sketch is very much like a storyboard that a game designer will produce when beginning a new game project. Just realize that nothing in the project is absolute at this point except for the requirements, and things will likely be changed as the process continues.

I envisioned the Pyboard mounted in a plastic case along with a quad AA battery pack, which should power it for many hours, assuming a low idle current draw. A bright LED would be mounted on a case side such that it would be visible to both players. The audio alert would be mounted in the case, with some holes drilled through the case to allow the sound to emit properly.

Two toggle switches would also be needed: one for power and the other to disable the audio alert.

The case itself would be attached to the net's top reinforced canvas strip using a steel plate mounted in the plastic case held in place by a very strong rare-earth magnet, essentially pinching the case to the canvas strip. This attachment method provides two benefits. The first is the attachment will not alter or damage the net in any fashion, which meets one of the project requirements. The second benefit is that using such a strong mechanical attachment will easily transfer any net vibrations to the Pyboard mounted in the case and then to the on-board accelerometer.

It is time to discuss the project sensor now that the initial prototype has been examined.

The Accelerometer

The Pyboard has a built-in accelerometer, which was used as the primary sensor to detect let ball occurrences. A small mechanical vibration is generated when the tennis ball just grazes the taut reinforced canvas strip that holds up the tennis net. The accelerometer is a sensor designed to convert vibrations into equivalent electrical signals. I believe it is important to explain how this happens so you will gain an appreciation and understanding of this sensor's operational capabilities as well as its constraints and limitations.

MMA7660FC Accelerometer

I will start by stating the MMA7660FC datasheet provided by the Freescale Semiconductor Corp. is the source of most of the following information. The company's formal name for this sensor is a 3-Axis Orientation/Motion Detection Sensor. I would strongly recommend that you download the datasheet PDF, which is available from the company's website, www.nxp.com. It is always a smart option to have as much information on hand as possible when you attempt to interface a device such as an accelerometer to a microcontroller. Figure 4-3 is a close-up macro-photograph of the accelerometer mounted on the Pyboard. The accelerometer is very small, only 3 × 3 × 0.9 mm in size.

Figure 4-3 *Pyboard accelerometer.*

This accelerometer is an example of an interesting technology known as micro-electromechanical systems (MEMS). MEMS devices are made up of semiconductor material with dimensions ranging from 0.02 to 1.00 mm, or 20 to 1000 micrometers. They are constructed using the same techniques as regular integrated circuits except they have actual mechanical moving components to implement the particular sensing or even actuating requirements. Believe it or not, MEMS electric motors have been created that can produce realistic forces to move things, albeit on a nano scale. Figure 4-4 is a micro-photograph of an electrostatic comb engine produced at Sandia National Labs. Truly, an amazing device.

Returning to the accelerometer, I think a quick refresher on what constitutes both acceleration and force should be useful for those readers who have been out of school for a while. Acceleration is defined as the rate of change of velocity. In mathematical terms, acceleration is defined as the time derivative of velocity, or stated as an equation:

$$a = dv/dt$$

The effects of acceleration are manifested on physical items by a force determined by Newton's second law of motion, which states the relationship that force is equal to the product of mass times acceleration, or in mathematical terms:

$$f = ma$$

This is the key relationship that allows the accelerometer to work. A force is induced in the moving part of the sensor, which in turn is the source of the electrical signal, which I will shortly discuss. The accelerations are the result of the vibrations generated when the tennis ball strikes or grazes the net.

This sensor uses a well-known principle of varying capacitance to measure acceleration. Figure 4-5 is a simplified diagram of one portion of the nano structure created within the sensor.

It is simply three plates made up of polycrystalline material. Two of the plates are fixed and surround the third plate, which is moveable and subject to the forces that result from any accelerations induced in the plate. These three plates are electrically equivalent to two series capacitors whose capacitance values are directly

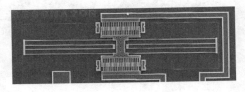

Figure 4-4   *Comb engine.*

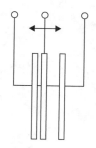

Figure 4-5   *Sensor nano structure.*

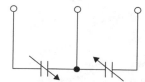

Figure 4-6   *Schematic for sensor series capacitors.*

related to the position of the moveable plate with respect to the fixed plates. Figure 4-6 is the electrical analog or schematic for these series capacitors.

There are electrical circuits in the accelerometer that convert the vary capacitance into equivalent acceleration values. Table 4-1 provides some key details regarding the performance and capabilities of the Freescale MMA7660FC accelerometer.

The six bits of digital output spread over a total axis acceleration range of 3 g's means that the g resolution is 3/64, or approximately 0.047 g per step, which should be fine for this application. The sensor's axes are orthogonal, meaning they are oriented at 90° to each other as shown in Figure 4-7.

Sensor Specification	Values/Descriptors
Acceleration range	Three axes, ±1.5 g each axis
Digital output protocol	I2C with six-bit resolution
Power requirements	2.4 to 3.6 V, 47 µA active mode
Motion detection	Acceleration, tilt, and gestures

Table 4-1   *Freescale MMA7660FC Key Specifications*

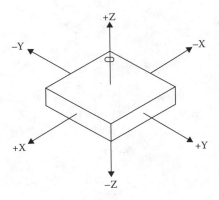

Figure 4-7 *Sensor axes orientation.*

The absolute sensor orientation really does not matter very much for this particular application, but it would be critical for a precision attitude application such as drone control.

Figure 4-8 is a portion of the complete Pyboard schematic, which shows the electrical interconnections for the accelerometer. It uses an inter-integrated circuit (I2C) protocol as mentioned earlier and previously introduced in Chapter 3.

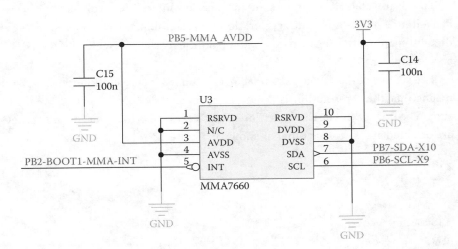

Figure 4-8 *Pyboard I2C interconnections to the accelerometer.*

You can clearly see from the figure that the sensor uses I2C bus 1, as the SCL pin is connected to X9 and the SDA pin is connected to X10 as detailed in Table 3-2. The sensor also uses two power supply inputs; the first, named AVDD, is used to power the sensor's analog circuitry, and the second, named DVDD, is used to power the digital circuits in the sensor. Sensor manufacturers often separate power supply inputs to minimize any possible noise interference generated in the analog circuits that might "leak" into the digital side. You can see from the figure that the DVDD input is tied permanently to the 3V3 supply (3.3 V), whereas the AVDD input is connected to a general-purpose input/output (GPIO) pin labeled PB5_MMA_AVDD. This means that this GPIO pin must be put into a high state in order to turn the sensor on. I must caution you that if you attempt to use a GPIO pin as a power source, you must ensure that the sensor load is within the allowable source power provided by the GPIO pin. In this case, this sensor only draws a very small load, as you can see from the power requirements detailed in Table 4-1. I will also demonstrate how to turn on the GPIO pin in the sample code.

There is another pin shown in Figure 4-8 for which I will provide a comment. This pin is labeled INT and is connected to the Pyboard's PB2-BOOT1-MMA-INT line. The INT pin is the sensor interrupts output, which may be triggered in one of seven modes as listed next:

- Front/Back
- Up/Down/Left/Right
- Tap
- GINT (real-time motion)
- Shake on X-axis
- Shake on Y-axis
- Shake on Z-axis

The particular interrupt mode selected depends on the value stored in the sensor's INTSU (0x06) register. If the GINT mode is selected, then the sensor motion can be accurately tracked based on the preset sample rate or how often the X-, Y-, and Z-axes values are updated. The sample rate ranges from 1 to 120 samples per second, making very precise motion tracking possible. I will not be using the interrupt mode in this application but will instead rely initially on polling the sensor to detect any let ball occurrences. This decision to use polling could change depending on whether or not the sensor is timely and responsive to the let ball strike.

At this point, I will use some low-level I2C class commands to demonstrate how this sensor interacts with the microcontroller using the I2C protocol. The actual application will use the pyb Acceleration class, which hides all the low-level I2C routines and permits you to develop at a higher abstract level. This hidden abstraction is one of the key features of MicroPython.

Low-Level I2C Command Demonstration

You will simply need to boot the Pyboard with an empty main.py, as I will be entering all the commands at the read-eval-print-loop (REPL) prompt. Next enter the commands as shown. I have provided comments that precede each command as to what it does and any expected display or response.

First you need to import the I2C class from the pyb library:

```
from pyb import I2C
```

Now you need to turn on the sensor as previously described:

```
accelPwr = pyb.Pin("MMA_AVDD")
accelPwr.value(1)
```

This command instantiates an I2C object named i2c using I2C bus 1:

```
i2c = I2C(1)
```

Direct the i2c object to scan for all active I2C devices. In this case, a hex 0x4C should be displayed.

```
i2c.scan()
```

The result of this command was [76], which is a single list element with a decimal value reflecting the discovered I2C device address. The hex equivalent of decimal 76 is 0x4C, which was the expected return value, indicating the sensor was turned on and successfully connected to the microcontroller over I2C bus 1.

A sensor register must be set to an appropriate value in order to read data from the sensor. Table 4-2 shows all the sensor registers along with their addresses and descriptive names.

The MODE register must be set to the active state in order to read sensor data. Table 4-3 shows the three modes, which can be set in the sensor along with the respective MODE register values.

The following command sets up the sensor to start generating x, y, and z data, which will be subsequently read:

```
i2c.mem_write(b"\x01", 0x4c, 7)
```

Hex Address	Name	Description
00	XOUT	Six-bit X output
01	YOUT	Six-bit Y output
02	ZOUT	Six-bit Z output
03	TILT	Tilt status
04	SRST	Sampling rate status
05	SPCNT	Sleep count
06	INTSU	Interrupt setup
07	MODE	Mode
08	SR	Various state setups
09	PDET	Tap detection
0A	PD	Tap debounce count
0B - 1F	Factory	Reserved

Table 4-2 *MMA7660FC Registers*

The three arguments in this command are:

b"\x01"—this is a numeric hex value of 1, which will be stored in the MODE register

0x4c—the I2C address for the sensor

7—the MODE register address

The sensor is now all set up for reading X-axis data after running the preceding command. This next command will return real X-axis acceleration values. All you need to do is set the Pyboard on a horizontal surface with the edge closest to the accelerometer pointing away from you. Gravity will do the rest of the work; no shaking or tapping needed, nor is it desired. Enter the following command to display the resting gravity force affecting the sensor:

```
i2c.mem_read(1, 0x4c, 0)
```

Mode	Standby	Test	Active
Hex value	00	04	01

Table 4-3 *MODE Register Values*

The byte hex value b\'01' was displayed, which indicates a very low level of gravity affecting the sensor in the X-axis based on its current orientation. Next, rotate the Pyboard 90 degrees such that the long edge that is closest to you is still touching the horizontal surface while the whole Pyboard is perpendicular to the surface. Repeat the previous command by simply pressing the UP ARROW key and then pressing ENTER. A value of b\'18' was displayed, which is 24 times greater than the value generated when the Pyboard was lying flat on the surface. This shows that gravity is having a maximal effect on the sensor's X-axis when oriented in this particular way. I am certain that the sensor's other axes would display identical readings if so tested.

Inclinometer

It is quite easy to create an inclinometer, which is also known as a tilt meter, using just the previous memory command. I recorded seven data points with the Pyboard tilted at angles ranging from 0 to 90 degrees in 15-degree increments. This data is shown in Table 4-4.

The data set was also charted as shown in Figure 4-9.

The chart line is a bit convex in shape, indicating that there is not a linear relationship between the tilt angle and the sensor's digital output. This is really not too surprising considering that there is only an absolute 24 numerical difference in the output for a total 90-degree tilt range, or stated another way, each output digit represents almost 4 degrees. This is a fairly coarse resolution, which is a direct result of the sensor's relatively low six-bit digital output. I only highlight this outcome to point out that it is always important to match any project requirements to a sensor's capabilities. If this project were to create a precision inclinometer, for example, then I would definitely choose a sensor with a higher resolution digital output of at least eight bits or even ten bits.

Tilt Angle (degrees)	Sensor Value (decimal)
0	1
15	8
30	13
45	17
60	20
75	23
90	25

Table 4-4   *Tilt Angle Data Set*

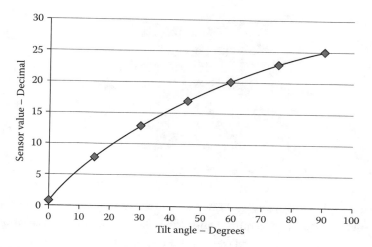

Figure 4-9 *Tilt angle chart.*

One last item I wish to mention is that the memory read command may easily be changed to directly display the tilt angle. The modified command is shown here:

```
(int.from_bytes(i2c.mem_read(1, 0x4c, 0)))*90/24
```

At this point, it would be appropriate to discuss the prototype hardware, as that will have an impact on the overall design and should be considered before launching into the software portion.

Hardware Design

The prototype hardware design is extremely simple, with most of the materials purchased at a local home improvement store. I used the sketch shown in Figure 4-2 as a guide and decided to use a shallow, plastic box as the main enclosure with a blank steel utility box cover to close up the plastic box, as shown in Figure 4-10. I did have to saw off two plastic tabs from the box, which were not needed for this design.

The steel cover is an important piece of this case, as that is how the case is held tightly to the net's canvas web strip. I also purchased a strong rare-earth magnet, which holds the case to the web strip as shown in Figure 4-11. Note that this was an early design photograph taken before I added the LED and switches to the case. I just wanted to ensure that the magnet holding idea would work before I proceeded too much further in the hardware design.

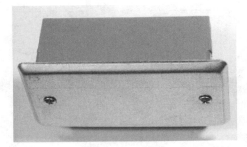

Figure 4-10 *Plastic case with steel cover.*

The case held very firmly, and I was convinced that it was a proper design that would meet the project requirements for a secure, nondamaging attachment device. I also confirmed that players on both sides of the net could easily see the LED array when it was attached.

The next step was to determine how to secure the Pyboard to the inside of the case. The obvious way was to use the two mounting tabs that are part of the Pyboard's printed circuit board (PCB). I just needed to be sure that the mounting would be rigid, as any net vibrations would have to be efficiently transferred through the mounts to the board and then to the sensor. Having a weak spot in this chain would seriously disrupt the mechanical vibration transfer. I elected to use relatively short nylon spacers along with 4-40 machine screws and nuts to secure the Pyboard to the case as shown in Figure 4-12.

Figure 4-11 *Case attached to a tennis net canvas web strip.*

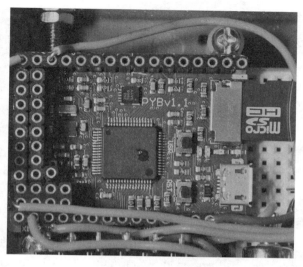

Figure 4-12 *Pyboard mounted to case interior.*

It is not especially critical where to locate the Pyboard within the case, other than it must be mounted on the plastic wall opposite the metal cover. Note that I also used small washers underneath the nuts to assist with the vibrational energy transfer.

There is one other item to consider: the size of the battery pack used to power the Pyboard, provided you chose to have all components mounted in the case. This was not a factor for my installation, as I elected to use a 6VDC sealed lead-acid battery power source connected to the Pyboard using a cable with a 2.1-mm barrel connector at one end. A matching socket was mounted on the case, and the Pyboard was connected to the power source through a toggle switch and wired directly to the VIN and GND pins as shown in the schematic in Figure 4-13.

There is an additional switch mounted next to the power switch, which controls the power to the Leaf audio module. After some consideration, I felt the most reliable design was to simply switch the power on or off to the module rather than do any software manipulations. The same signal that directs the LED module array to turn on also directs the audio module to turn on. However, if no power is being supplied to the audio module, then no sound is being created. Figure 4-14 is the schematic showing the LED and audio modules connections to the Pyboard.

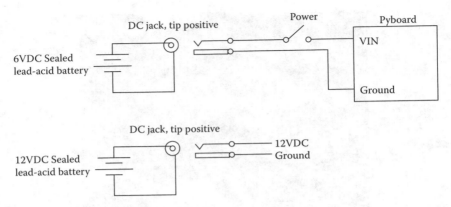

Figure 4-13   *Power supply schematic.*

Note that I used an NPN driver transistor for the LED driver circuit because the LED array current load exceeded the rating for a single GPIO pin. I used a 2N3904 transistor, but just about any general-purpose NPN switching transistor will suffice. The audio module is separately powered, and all the GPIO pin does is provide a signal level, which has a negligible current draw. Figure 4-15 shows the case interior and exterior with all components and modules mounted.

The LED module is mounted to the top of the case such that the players can easily see it. Figure 4-16 shows the LED module mounted using two 4-40 ½-inch machine screws and nuts.

The LED module also comes with an adhesive strip attached, which you may use in lieu of mounting with the machine screws. You will, however, still need to drill a hole into the side of the case for the two LED power leads. Incidentally, the module already has a built-in current-limiting resistor, so you will not have to worry about that. Just control it with 12VDC through the NPN switching transistor as shown in the schematic in Figure 4-14.

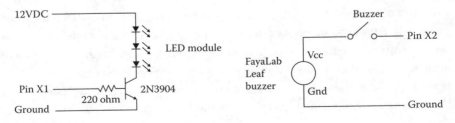

Figure 4-14   *LED and audio modules schematic.*

Figure 4-15 *Fully assembled case with all modules mounted.*

The two switches and power connection sockets are shown in Figure 4-17.

Note that I labeled one of the power sockets as 5VDC but in reality, I was using a 6VDC sealed lead-acid battery to power the Pyboard. You might be concerned that the two power connectors are identical and you could inadvertently plug the 12VDC supply into the socket marked 5VDC. If that were to happen, the voltage regulator on the Pyboard has an absolute maximum input voltage rating of 12VDC and nothing bad should happen. However, I used a 12VDC-rated sealed lead-acid battery for the 12VDC supply, and as most readers know, a fully charged 12VDC battery can provide more than 14VDC, which unfortunately exceeds the voltage regulator limit. So my strong advice is to double-check that you are plugging in the appropriate power supplies into the correct sockets or change one of the sockets to prevent this potential problem.

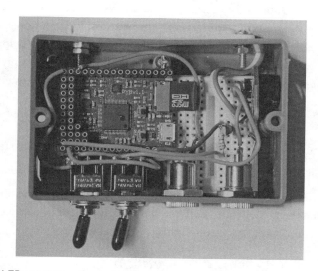

Figure 4-16 *LED module mounted to case top.*

Figure 4-17 *Switches and power sockets.*

The one final hardware configuration was to ensure that the players could hear the audio alert tone. The plastic case I used already had some slits in the case, which would allow the tone to be heard. Just ensure you have some neatly drilled holes if the case you use is completely enclosed.

It is now time to create a program that will detect the net vibrations to meet project requirements. I will be using the Accelerometer class, as mentioned previously, to program the sensor.

Main Program

This program will initially be designed to sense all three axes to provide the maximum amount of data to be used in detecting a let ball strike. It may turn out that the data for one or two axes is not really necessary, but at the beginning stage that is not a certainty. A project program should always be developed in stages, where functions to meet certain requirements are first developed and tested and then others are gradually incorporated into the program. Such a staged approach usually minimizes program debugging and has been shown, at least in my experience, to be the fastest and most productive way to create a project program. Only a few stages are involved in this program:

- Sense three axes for accelerations
- Determine an appropriate data set representative of a let ball strike
- Turn on visual and optional audio alerts upon let ball detection

As you probably can see, more than pure coding necessary is to complete each stage. Writing the actual code is often a minor component of completing a stage, as you will see as I progress through this portion of the project.

Sense Three Axes of Acceleration

The quickest and easiest way to start reading X, Y, and Z acceleration values is to enter a script at the REPL prompt, which will instantiate an Accel object and then

```
MicroPython v1.5.1-66-g66b9682 on 2015-12-04; PYBv1.1 with STM32F405RG
Type "help()" for more information.
>>> import time
>>> accel = pyb.Accel()
>>> for i in range(20):
...      print(accel.x(), accel.y(), accel.z())
...      time.sleep(1)
...
4 -1 22
4 0 21
2 0 22
10 12 2
-13 14 -2
13 14 14
28 -6 31
-32 -5 19
-17 -2 11
-9 2 -5
-25 -5 10
-14 1 -7
-24 -2 16
-15 1 -18
-17 -4 -1
-13 0 -3
-28 -4 18
-31 -13 6
-4 6 -21
-14 0 13
>>> ▮
```

Figure 4-18 *Acceleration script and set of sample readings.*

take a set of 20 readings. Figure 4-18 is a screen shot of both the script and a sample set of readings.

The time.sleep(1) statement allowed plenty of time to shake the Pyboard to create different readings in all three axes. The raw values will be in the range of –32 to +31, which are in decimal format and already converted from the byte hex values that were displayed using the low-level I2C commands. I did not have to violently shake the Pyboard to obtain this set of readings, which indicated to me that tennis net vibrations should be able to be detected with ease. I also want to point out how simple it is to obtain an acceleration reading using the Accel class. I just instantiated an object named accel and then used the x(), y(), and z() methods to return an acceleration values from their respective axis. All the low-level details such as setting the sensor address, configuring the mode, and doing memory reads are completely hidden and done for you automatically. As I previously stated, MicroPython is a powerful language and allows you to focus on embedded application development and not get bogged down in the support details.

I will now discuss the next stage, having demonstrated how easy it is to acquire the acceleration raw data.

Determine an Appropriate Data Set Representative of a Let Ball Strike

This stage is somewhat more difficult to master than the previous one. This stage's goal is to determine a set of X, Y, and Z accelerations, which happen when a tennis ball just grazes the top canvas web support strip. The problem that naturally arises is how to differentiate a let ball strike from all the other vibrations that a tennis net is constantly creating, such as from player movements, wind, balls hitting into the net, etc. After some consideration, I decided to create a simple data logger to record the different vibrations, including let ball strikes, and then use the recorded data to help me establish a specific set of readings that only happen with a let ball strike. Fortunately, the Pyboard has such a data recorder onboard in the form of an SD card. All that I needed to do was write a program to record data and then correlate that data with the actual event that it was happening when the record was made. The next section deals with how I set up a data logger, which used an SD card to record the vibration data.

SD Data Recorder

The following is a stand-alone data logger program I wrote that logs on to a micro SD card the vibrations resulting from various tennis ball strikes. This was done in order to obtain sufficient data to establish good criteria as to when an actual let ball strike happens.

```python
# datalogger.py

# Logs the data from the accelerometer to a file on the SD-card

import pyb

accel = pyb.Accel()

blue = pyb.LED(4)

f = open('test.dat', 'w')

f.write('Tennis net vibration record\n')

f.write('Three full on net hits\n')

f.write('Five grazes\n')

blue.on()

for i in range(120):
```

```
    t = pyb.millis()

    x,y,z = accel.filtered_xyz()

    f.write('{},{},{},{}\n'.format(t,x,y,z))

    pyb.delay(1000)

f.close()

blue.off()
```

I found that I also had to boot the Pyboard into the non–mass media storage mode, where it does not present a virtual USB drive to the host computer. It acts instead as a human interface device (HID), similar to a mouse or keyboard. It is simple to change the boot configuration; however, you must load the data logger program first, as you have no access to the main.py file once you change the boot. Follow these steps, which will allow you to load the data logger program into the micro SD card and change the boot configuration:

1. Format a blank micro SD card on a host computer. I believe most new micro SD cards already come formatted, so this step is not likely to be required.

2. Insert the micro SD card into the Pyboard and then plug the USB cable into the host computer.

3. A virtual USB drive should appear on the host with some default name already assigned to the card. If not, it might be "NO NAME," as was the case with my MacBook Pro.

4. Open the USB drive and confirm that there are at least two files named boot.py and main.py in the root folder.

5. Copy the datalogger.py program listed earlier into the main.py file.

6. Uncomment the line #pyb.usb_mode('CDC+HID') # that acts as a serial device and a mouse that appears in the boot.py file.

7. Eject the USB drive from the host and unplug the USB cable.

The Pyboard is now all set to record 120 seconds of tennis net vibrations to a micro SD card whenever power is applied to it. The onboard blue LED will also stay on for the entire recording time to provide a visual clue that data is being acquired. However, that LED cannot be seen when the Pyboard is mounted in the plastic case and positioned on the net.

Experimenting on the Tennis Court

This was the fun part of the project where I went to a local tennis court and attached the module to the net's supporting canvas strip. I then had two minutes to hit tennis balls into and at the top of the net to record representative vibrations. Let me state that I am in no way close to being even a good tennis player; however, my years playing racquetball did help me with this experiment. I actually repeated the data acquisition several times before I felt that I had a reasonable data set with which I could set the let ball criteria. Figure 4-19 is a portion of the micro SD card data file that I finally used.

The criteria were encoded into a compound if statement that continuously checked the filtered X, Y, and Z outputs for averaged values that fell within the expected ranges. I did notice that the X values were relatively constant and independent of what was happening to the net, so I elected to ignore them in the detection algorithm. Figure 4-19 also shows the acceleration values I selected, which directly related to the actual ball strikes on the net.

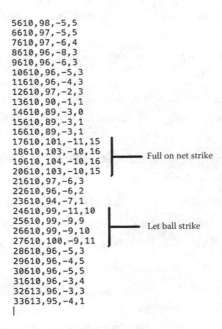

Figure 4-19   *Portion of the experimental data set.*

Another concern I had was the variability in the velocity of the ball striking the net. I am certain that my efforts were probably producing velocities and corresponding accelerations that were probably half the speed of what a professional player would generate. I decided that I would use a ratio test that would remove the velocity variability in the let ball detection test. The test simply checks the ratio of the Y to Z accelerations, which makes the absolute acceleration values irrelevant. The following compound if statement shows how this check is implemented.

```
x,y,z = accel.filtered_xyz()
if y < -5: if z/abs(y) < 1.25:
    <do something>
```

All that was left was to finish the rest of the code once I had a good set of criteria to recognize a let ball strike.

Turn On Visual and Optional Audio Alerts upon Let Ball Detection

Activating the visual and audio alerts is fairly simple and only requires setting two GPIO pins high for a few seconds and then low again. The following code snippet shows how this is done:

```
import pyb
. . .
pinX1 = Pin('X1', Pin.OUT_PP) # LED control
pinX2 = Pin('X2', Pin.OUT_PP) # buzzer control
. . .
while True:
. . .
    # turns on pins X1 and X2 for 5 seconds
    pinX1.high()
    pinX2.high()
    pyb.delay(5000)  # 5000 millisecs
    pinX1.low()
    pinX2.low()
. . .
```

The LED and audio alert circuits must be connected to pins X1 and X2 as previously discussed.

This completes the stage discussions, and all that's left is to show you the final program.

Final Program

The following listing is the final program that must be entered into the main.py file located on the micro SD card. The boot configuration can be left in the HID mode, as it will make no difference in the Pyboard's operation because it will be completely battery operated.

```
import pyb
import math
pinX1 = Pin('X1', Pin.OUT_PP)
pinX2 = Pin('X2', Pin.OUT_PP)
accel = pyb.Accel()
while True:
    x,y,z = accel.filtered_xyz()
    if y < -5: if z/abs(y) < 1.25:
        # turns on pins X1 and X2 for 5 seconds
        pinX1.high()
        pinX2.high()
        pyb.delay(5000)   # 5000 millisecs
        pinX1.low()
        pinX2.low()
```

You should copy this code and put it into the SD card's main.py file. The Pyboard is already set up to run the program code from the SD card, and you will avoid the issue of trying to plug in a micro USB cable to the mounted Pyboard. In fact, I strongly urge you not plug in a cable because it just gets cramped in the case and it is so easy to inadvertently cause a short. I know, because it happened to me and I unfortunately ruined my Pyboard. A word to the wise.

I did get a chance to test the whole circuit before I destroyed the Pyboard and found it to function as expected, with the LED module activated with a slight shaking of the case. I have no doubt it will function quite well attached to a real tennis net. My only comment would be that the detection algorithm might have to be tweaked a bit if too many false alarms are generated. That should not be too much of a problem, however, as it will only entail changing the Y-axis and the ratio values.

Summary

This chapter's project was about designing and building a let ball detector. A let ball is a tennis term for a served tennis ball, which just grazes the top of the net and subsequently lands in the opponent's correct service area. It is a "do over" situation without a fault assessed to the server. Let balls are often hard to detect without a referee present, which is the reason for this automated device. The detector takes advantage of the Pyboard's three-axis accelerometer as a primary sensor.

In this project, I took you through all my normal steps for project design and build, including writing all the requirements, building a prototype, and finally modifying it based on initial test results.

A section on how the three-axis accelerometer functions was also included, as it is very important to understand how a sensor works in addition to simply knowing how to use it. I also covered some I2C interface considerations, which allow the digital accelerometer sensor to communicate with the Pyboard's processor.

The chapter finished with a thorough discussion on how to design a program to implement the let ball occurrence. I detailed how to generate calibration data, which is critical for the detector's proper operation.

5

LCD and
Touch-Sensor Board

In this chapter I will show you how to install and program the liquid crystal display (LCD) and touch-sensor skin (board) available from the MicroPython store. This board is shown in Figure 5-1 and comes with two 8-pin male headers that must be soldered onto the board in order for it to be plugged into the Pyboard.

The LCD and touch-sensor board is a compact module that is designed to plug directly on top of the Pyboard. It has a 128 × 32 pixel, monochrome graphical LCD, as well as a capacitive touch-sensor chip with 12 built-in channels. I will discuss in later sections how these sensors operate, as it is interesting technology and important for you to understand when building modern microcontroller projects. But first, I will present some key specifications regarding the LCD board.

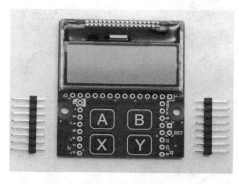

Figure 5-1 *LCD and touch-sensor board.*

LCD Board Specifications

The monochrome LCD screen is 128 × 32 pixels and has an integral white backlight that may be turned on and off by software. The touch-sensor chip is a Freescale chip—model MPR121—that supports up to 12 channel-capacitive touch sensors. Four physical capacitive touch pads are part of the module's printed circuit board (PCB) and labeled A, B, X, and Y. There are also 12 soldering pads, which expose all 12 input sensor channels. These soldering pads may be used to connect up to eight additional touch sensors, if desired. These pads also allow direct access to the MPR121 pins in a general-purpose input/output (GPIO) mode, if the pins are individually not configured as touch-sensor inputs. I will discuss the GPIO operational mode more in a later section. The MPR121 chip connects with the Pyboard using an inter-integrated circuit (I2C) channel, which was described in an earlier chapter. You should also note that the LCD screen itself is not a touch screen, just the four touch pads located below, as may be clearly seen in Figure 5-1.

Figure 5-2 shows the back of the module where you may clearly see the MPR121 surface-mounted chip.

There is another key controller/driver chip on the board, which is not visible in either Figure 5-1 or Figure 5-2. That chip is the Sitronix ST7565R, a 132 × 65 dot matrix LCD controller/driver, and it is totally integrated into the LCD. The LCD itself is manufactured by New Haven Display International and is designated as model NHD-C12832A1Z-FSW-FBW-3V3. It has 128 × 32 transflective pixels

Figure 5-2 *Back side of the LCD board.*

with a side white LED backlight. The LCD obviously uses less than half of the Sitronix controller's pixel capability. The characters on the display are visible without the backlight on, but they are much easier to read with it operating. Individual pixel dots are almost invisible without the use of the backlight. This will become readily apparent in a later graphics demonstration.

The Sitronix controller communicates with the Pyboard using the I2C bit-serial interface, which I discussed in an earlier chapter. It is an interesting situation where a single add-on module uses two distinctly different communications protocols for interoperability with the Pyboard. The MicroPython language makes using these protocols seamless and transparent and allows the developer to focus on the application, not on driver peculiarities.

Initial LCD Module Operations

The LCD module will first need to be mounted to the female Pyboard headers as shown in Figure 5-3. This is the X configuration, which will be explained shortly.

Ensure that the LCD module pin with a white block surrounding it plugs directly into the matching pin with a white block on the Pyboard. The left-side

Figure 5-3 *LCD module mounted to the Pyboard.*

mounting holes on the LCD module and the Pyboard should also align concentrically with each other. Figure 5-4 is a side view showing these white block pins to help you properly mount the LCD module. Mismatching the pins could have dire consequences for both the LCD module and the Pyboard.

The following commands will verify that the LCD module display functions properly with the Pyboard. After mounting the LCD module, you should connect the Pyboard to the host computer and open a terminal command window. I used my MacBook Pro as a host, as was done in previous chapters, and opened a terminal window to enter the Python commands. To start displaying characters on the LCD screen, you must instantiate an LCD object using the following command:

```
lcd = pyb.LCD('X')
```

This command creates an object named lcd of the LCD class, which in turn is part of the pyb library. The X argument in the LCD class refers to the display orientation. The display also has a Y argument that works when the LCD module is plugged into the upper section of the Pyboard's side header sockets, essentially the exact opposite of the X configuration. Figure 5-5 shows the LCD mounted in the Y configuration. The Y configuration allows the AMP skin to be plugged in simultaneously with the LCD module. I will be discussing the AMP skin in the next chapter. From this point on, I will be using the X configuration and its respective mounting for all the demonstrations in this chapter.

Figure 5-4   *Side view of LCD module mounted to the Pyboard.*

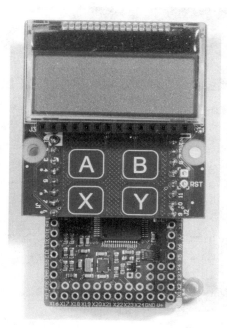

Figure 5-5 *LCD Y mount position.*

Once the `lcd` object is instantiated as discussed earlier, you can now call methods contained in the class. The first method you should call is the one that turns on the white LED backlight. This command accomplishes it:

```
lcd.light(True)
```

The display should now be brightly lit after executing this command. I also could not find any way to dim the display using software commands, so you will have to be satisfied with it being fully on or off.

Now it is useful to display some characters. As always, the traditional first set of characters should be the "Hello World" message. Enter the following to see that displayed:

```
lcd.write('Hello World!')
```

Note that single quotes delimit or surround the text characters to be displayed. You may also include escaped characters, such as newline, to position the cursor for the next line of text. The following shows how to enter four lines of text:

```
lcd.write('The first line\nThis is the 2nd\nThe third line\nNow the last')
```

Figure 5-6 shows the LCD that results from this command.

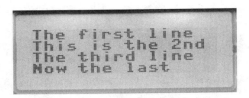

Figure 5-6 *Multiline LCD.*

The LCD is restricted to a 4 × 16 character display based on the internal font set used by the LCD controller. Going beyond 16 characters on a single line will cause the controller to automatically insert carriage return/line feed characters to force a new line. You must be aware of this limitation or you will definitely mess up your text displays.

In Table 5-1, I have listed all the methods, including the constructor, which are contained in the LCD class.

Method Name	Description
LCD(display orientation)	This is the constructor. The only acceptable orientation arguments are X or Y.
lcd.command(instruction_or_data, buffer)	An instruction or data byte can be sent directly to the LCD controller. The first argument is set to 0 for an instruction and 1 for data. The second argument is the actual instruction or data in a buffer structure.
lcd.contrast(value)	Sets the display contrast. Acceptable values range from 0 to 47.
lcd.fill(color)	Fill the hidden buffer with a color with a value of either 0 for white or 1 for black. Use show() to display the hidden buffer.
lcd.get(x, y)	Return the pixel value (either 0 or 1) at the (x, y) coordinates.
lcd.light(value)	Turn the backlight on (value = True or 1) or off (value = False or 0).
lcd.pixel(x, y, color)	Set a specified pixel color (0 or 1) at the given (x, y) coordinates. Writes to the hidden buffer. Use show() to display the hidden buffer.
lcd.show()	Display or show the hidden buffer.
lcd.text(str, x, y, color)	Draw the text contained in the string starting at the (x, y) coordinates with a color of 0 or 1. This writes to the hidden buffer. Use show() to display it.
lcd.write(str)	Immediately display the string starting at the current cursor position.

Table 5-1 *LCD Methods*

You should note that the Sitronix ST7565R LCD controller actually supports many more instructions than are shown in the table. Most of those instructions are not applicable for this write-only application; however, I will demonstrate one interesting and useful instruction in a later section.

It is now time to show you some simple graphics operations that can be done with the LCD now that you know how to display straightforward text.

LCD Graphics Demonstration

This demonstration requires that the following program be entered into the main.py file located in the PYFLASH directory, which appears when you connect the Pyboard to a host computer. This code comes from the micropython.org LCD tutorial. I have included some comments to help clarify what is happening in the code.

```
import pyb
x = y = 0
dx = dy = 1

lcd = pyb.LCD('X') # instantiate an lcd object
lcd.light(True)    # turn on the backlight
lcd.contrast(47)   # helps make the dot visible

while True:
# update the dot's position
x += dx
y += dy

# make the dot bounce off the screen edges
if x <= 0 or x >= 127: dx = -dx
if y <= 0 or y >= 31: dy = -dy

lcd.fill(0)          # clears the hidden buffer
lcd.pixel(x, y, 1)   # paint a dot
lcd.show()           # show the hidden buffer
     pyb.delay(75)      # pause for 75 ms
```

This program should automatically run when you restart or reset the Pyboard. It simply causes a dot to continually zigzag across the 128 × 32 LCD field, reversing as it encounters a boundary edge, similar to a "pong" display of yesteryear. Figure 5-7 is a close-up snapshot of the dot in progress.

You will immediately notice a series of dots in the figure resembling a "micro-comet" instead of a single dot. These dots are not an artifact of the camera taking

Figure 5-7 *Running graphics program display.*

the photograph, but are due to the LCD response time. LCD pixels take a finite time to both turn on and turn off, which in the case for this particular LCD screen ranges from 150 to 250 ms. The dot is programmed to move one diagonal step every 75 ms, which means that at least the previous three dots will also be visible as well as the current dot being displayed. The preceding dots will also be diminishing in intensity. Interestingly, if you examine Figure 5-7 closely, you should see that the lead, or current, dot is not as intense as its immediate predecessor. This is also due to the transient response time it takes to build to full intensity. I should note that technically the dots are not being displayed as would happen with an LED pixel, but instead the LCD pixel polarization is being flipped to make a dark dot appear at the desired coordinates. This polarization "flipping" is the reason why LCDs have been considered "slow" displays compared with LED displays. Sometimes this slowness has been called motion blurriness, especially when it comes to TV applications. However, modern LCD TVs have specially constructed displays and associated circuits, which greatly minimize motion blur or, more appropriately, transient response times.

A bit later in the chapter I will present another graphics example that works with the built-in touch-pad sensors. Understanding the previous demonstration and the upcoming touch-pad example should provide you with good ideas on how to use graphics in addition to regular text displays with this LCD module.

I will next demonstrate how to enter an external command to the LCD controller as I had mentioned earlier.

Using an External Command with the LCD Controller

The Sitronix ST7565R LCD controller is capable of handing additional instructions beyond what is shown in Table 5-1. The `lcd.command(instruction_or_data, buffer)` is the means by which these additional instructions can be sent to the controller.

The Sitronix datasheet lists 22 instructions, but unfortunately, only a few are applicable for this application given the constraints of the write-only, 132 × 32 pixel LCD. However, one instruction is quite useful, which reverses the display present, making all the light pixels dark and vice versa. I used the same lcd write instruction shown earlier, which writes four lines to the display:

```
lcd.write('The first line\nThis is the 2nd\nThe third line\nNow the last')
```

After this is run, enter the following:

```
s = '\xa'
lcd.command(0, s)
```

The display should reverse immediately after entering the lcd.command instruction. Figure 5-8 shows the reversed display.

Enter the following to restore the normal black text on a white background display:

```
s = '\xa6'
lcd.command(0, s)
```

I also tried experimenting with some of the other Sitronix instructions but found that they were either ineffective due to the write-only circuit or did nothing that could not be accomplished using the LCD class methods. However, feel free to experiment to see if you can achieve some results not available using the class methods. The worse you can do is lock up the display, which is easily undone by a power-off reset.

This section concludes my initial demonstration of how to use the LCD portion of the LCD module. It is time to turn our attention to the touch controller.

Touch Controller

The LCD module has a 12-channel touch controller as mentioned in the chapter introduction. This controller uses the Freescale MPR121 to sense capacitive changes that happen on any one of the 12 input sensor channels. The LCD board has four touch-sensor pads connected to four sensor inputs as shown in Figure 5-1. The next section discusses how capacitive touch sensing functions, which is in line with my thinking that it is important for users to be aware of how technology works in addition to just knowing how to use it.

Capacitive Sensing

I will start by stating that much of this material is based on the Freescale application note AN3889, which is a thorough discussion of capacitive sensing technology. This note is freely available for download, and I urge interested readers to refer to it to gain a further understanding of the technology beyond my brief introduction.

The basic physical device involved in this sensor is the capacitor, which essentially can be thought of as two parallel metal plates in close proximity to each other and separated by a dielectric substance. Electrical charge is stored in the dielectric substance whenever a voltage (or, more formally, voltage potential) is applied to the plates. An electrical current flow will commence between the plates whenever the applied voltage changes with time. This behavior is neatly captured by this classic equation relating voltage and current in a capacitor.

$$\frac{dV}{dt} = \frac{I}{C}$$

The voltage appearing between the capacitor plates summed over a specified time interval T is

$$V = \frac{I * T}{C}$$

This capacitor voltage buildup can be seen in the charging diagram in Figure 5-9, which is taken from the application note.

The peak voltage shown in the figure is inversely proportional to the capacitance value, C, meaning a higher value for the total capacitance will produce a lower peak value. The MPR121 chip has a ten-bit multiplexed analog-to-digital converter (ADC), which can accurately measure the peak voltage. Any small additional capacitance added to the touch pads' baseline capacitance will definitely be measured. This additional capacitance could be a human finger or even a stylus touching the sensor pad. The peak voltage on each enabled touch-sensor input is periodically sampled, and if a lower voltage is detected that is lower than the baseline peak voltage, this would indicate that something or someone has touched that sensor.

Figure 5-8   *Reversed display.*

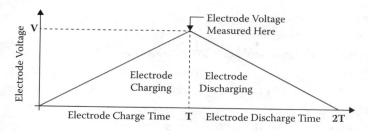

Figure 5-9   *Charging diagram.*

I will next show you a portion of the LCD module schematic concerning the touch-sensor chip and the touch pads.

LCD Module Touch-Sensor Schematic and MPR121 Registers

Figure 5-10 shows the touch-sensor schematic along with the four touch pads included within the LCD module.

You should be able to see that the four touch pads, Y, X, B, and A, are connected to ELE0 to ELE3, respectively. In addition, all the touch inputs ELE0 to ELE11 are brought out to the solder pads on J3, pins 3 to 14, respectively. Notice that the sensor inputs from the MPR121 chip are labeled ELE0 to ELE3 and LED0. LED0/ELE4 to LED7/ELE11. This peculiar labeling is used to indicate that the first four inputs are solely for touch pad sensing, while the remaining eight could be used for touch pad or GPIO use. I will show you how to use some of these uncommitted pins in a GPIO demonstration later in the chapter. Incidentally, pins 1 and 2 on J3 provide both ground and 3.3-V connections, which makes it quite convenient for GPIO experiments.

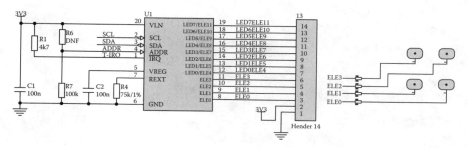

Figure 5-10   *Touch-sensor schematic.*

The MPR121 chip contains 128 eight-bit registers, which are used to both configure the chip and provide the means through which data and status can be reported back to the Pyboard. Table 5-2 provides details for a small representative sample of these registers for your information. It is not too critical that you know how to directly use most of these registers, as all that tedious work is done for you by the MPR121 driver software that I will discuss a bit later. I will be using some of the GPIO registers for the GPIO demonstration as mentioned earlier. All the details regarding these registers were culled and summarized from the Freescale MPR121 datasheet.

Note the reference to ELEPROX in the register listing. This refers to a new feature for the MPR121 chip where all the touch-sensor inputs are connected in parallel so as to implement a proximity sensor. This "virtual" sensor would trigger when a human hand approaches but does not touch the sensor array. The datasheet does not specify how close the hand must be to the array, as that parameter probably depends on the size of each individual touch sensor. I will not be experimenting with this feature; however, I did want to make note of it for those readers interested in this new feature.

Register Description	Address (hex)
ELE0 – ELE7 Status	0x00
ELE8 – ELE11, ELEPROX status	0x01
ELE0 Baseline Value	0x1e
ELEPROX Value	0x2a
ELE0 Touch Threshold	0x41
ELE0 Release Threshold	0x42
ELE0 Electrode Current	0x5f
ELE0, ELE1 Charge Time	0x6c
GPIO Control Register 0	0x73
GPIO Control Register 1	0x74
GPIO Data Register	0x75
GPIO Direction Register	0x76
GPIO Enable Register	0x77
GPIO Data Set Register	0x78
GPIO Data Clear Register	0x79
GPIO Data Toggle	0x7a
Soft Reset Register	0x80

Table 5-2 *Representative Sample of the MPR121 Registers*

In the next section, I discuss the MPR121 support software, which allows the touch sensor to function very well with the Pyboard.

MPR121 Driver Software

You may be a bit surprised that the pyb library does not contain a touch-sensor class. What it does contain is the I2C class, which directly supports the I2C bit-serial protocol, which the MPR121 uses to communicate with a microcontroller such as the Pyboard. Wrapper software is therefore needed, which uses the I2C class to directly implement the touch sensor commands required by the MPR121 chip. Fortunately, the good folks at www.miropython.org have provided this software in the form of a MicroPython file appropriately named MPR121.py. This driver is directly available from the touch-sensor tutorial web page, https://micropython.org/doc/tut-lcd-skin. I have also included the full listing next to which I have some additional comments to help clarify what is happening within the code.

```
"""
Driver for the MPR121 capacitive touch sensor.
This chip is on the LCD32MKv1 skin.
"""
import pyb
# register definitions
TOUCH_STATUS = const(0x00)
ELE0_FILT_DATA = const(0x04)
ELE0_TOUCH_THRESH = const(0x41) # this is the peak baseline voltage
DEBOUNCE = const(0x5b)
ELEC_CONFIG = const(0x5e)

class MPR121:
    def __init__(self, i2c):
        self.i2c = i2c # i2c is the I2C object from the main.py code
        self.addr = 90 # I2C address of the MPR121

        # enable ELE0 - ELE3, first four sensor inputs - see Figure 5-9
        self.enable_elec(4)

    def enable_elec(self, n):
        """Enable the first n electrodes."""
        self.i2c.mem_write(n & 0xf, self.addr, ELEC_CONFIG)

    def threshold(self, elec, touch, release):
        """
        Set touch/release threshold for an electrode.
        Eg threshold(0, 12, 6).
        """
        buf = bytearray((touch, release))
        self.i2c.mem_write(buf, self.addr, ELE0_TOUCH_THRESH + 2 * elec)
```

```
def debounce(self, touch, release):
    """
    Set touch/release debounce count for all electrodes.
    Eg debounce(3, 3).
    """
    self.i2c.mem_write((touch & 7) | (release & 7) << 4, self.addr, DEBOUNCE)

def touch_status(self, elec=None):
    """Get the touch status of an electrode (or all electrodes)."""
    status = self.i2c.mem_read(2, self.addr, TOUCH_STATUS)
    status = status[0] | status[1] << 8
    if elec is None:
        return status
    else:
        return status & (1 << elec) != 0

def elec_voltage(self, elec):
    """Get the voltage on an electrode."""
    data = self.i2c.mem_read(2, self.addr, ELE0_FILT_DATA + 2 * elec)
    return data[0] | data[1] << 8
```

This driver code is in the form of a class definition, which means it must be instantiated and used within an application program, which is the subject of the next section regarding the initial touch-sensor test.

Initial Touch-Sensor Test

You can initially test the touch sensor in the read-eval-print-loop (REPL) mode by directly using the I2C class methods to communicate with the MPR121 chip without the need for the MPR121.py driver code. All you need is the I2C address for the MPR121 chip, which is 0x90. Enter the following at the MicroPython prompt:

```
i2c = pyb.I2C(1, pyb.I2C.MASTER)
i2c.mem_write(4, 90, 0x5e)
touch = i2c.mem_read(1, 90, 0 )[0]
```

The first command instantiates an I2C class object named i2c. The second command enables the first four touch-sensor input lines, ELE0 to ELE3. And the last command reads the states of the touch pads A, B, X, and Y and then stores them in the touch variable. Figure 5-11 is a screen shot of the terminal window after I entered the initial two commands and then executed the third one while touching the A pad. I then entered the touch variable at the prompt to see its value, which was 8. I repeated the read command three more times and then read out the values, which were 4 for B, 2 for X, and finally, 1 for Y.

```
● ● ●    ⌂ donnorris — screen /dev/tty.usbmodem1412 ▸ SCREEN — 80×24
MicroPython v1.5.1-66-g66b9682 on 2015-12-04; PYBv1.1 with STM32F405RG
Type "help()" for more information.
>>> i2c = pyb.I2C(1, pyb.I2C.MASTER)
>>> i2c.mem_write(4, 90, 0x5e)
>>> touch = i2c.mem_read(1, 90, 0)[0]
>>> touch
8
>>> touch = i2c.mem_read(1, 90, 0)[0]
>>> touch
4
>>> touch = i2c.mem_read(1, 90, 0)[0]
>>> touch
2
>>> touch = i2c.mem_read(1, 90, 0)[0]
>>> touch
1
>>> touch = i2c.mem_read(1, 90, 0)[0]
>>> touch
12
>>> ▌
```

Figure 5-11 *Terminal window for initial touch test.*

Touching two pads simultaneously results in the sum of the individual pads. I touched the A and B pads together and observed the value of 12, as expected, and which is also shown in the figure.

After you complete the initial REPL check, you should load the mpr121.py file in the PYBFLASH USB drive and then store the following code in the main.py file:

```
import pyb
import mpr121
m = mpr121.MPR121(pyb.I2C(1, pyb.I2C.MASTER))
for i in range(100):
    print(m.touch_status())
    pyb.delay(1000)
```

Figure 5-12 shows the result of running this short program and my touching the various touch pads with a stylus two times.

The next program advances the previous one by having each touch pad directly control a corresponding LED.

LEDs Controlled by Touch Pads

This demonstration will show you how to use the LCD module touch pads to turn LEDs on or off, depending on which pad is activated. A set of four LEDs must be connected to the Pyboard in order for this demonstration to work. Figure 5-13 is a Fritzing diagram that shows how to connect the four LEDs along with four current-limiting resistors to the Pyboard.

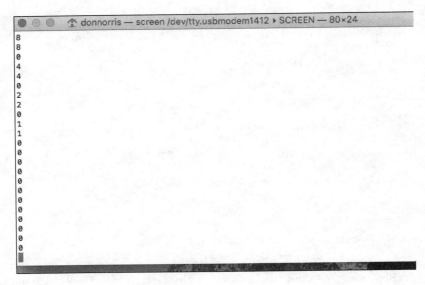

Figure 5-12 *Program results.*

Note that the LCD module is not shown on the diagram, as there are no additional connections to the module that are required for this demonstration. I chose to use the Pyboard pins Y1 to Y4, as they are readily accessible compared with the Xn pins, which are already either in use by or covered by the LCD module. I have

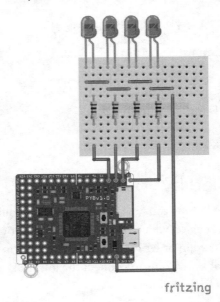

Figure 5-13 *Fritzing diagram showing LED connections to the Pyboard.*

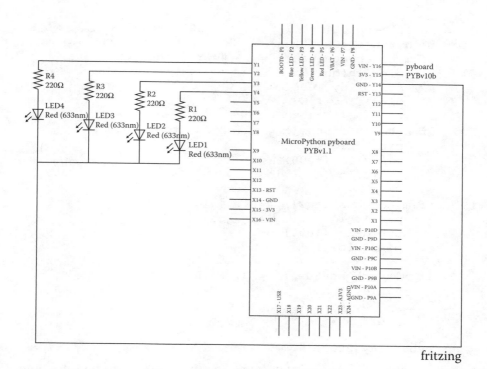

fritzing

Figure 5-14 *LED demonstration circuit schematic.*

also included Figure 5-14, which is an electrical schematic for the LED demonstration circuit. It should help you build the circuit without too much difficulty.

The next step you need to take is to put the following program into the main. py file. Note that this program also needs the mpr121.py file to be in the PYB-FLASH USB drive in order to work.

```
import pyb
from pyb import Pin
import mpr121

pinY1 = Pin('Y1', Pin.OUT_PP)
pinY2 = Pin('Y2', Pin.OUT_PP)
pinY3 = Pin('Y3', Pin.OUT_PP)
pinY4 = Pin('Y4', Pin.OUT_PP)
```

```
m = mpr121.MPR121(pyb.I2C(1, pyb.I2C.MASTER))
for i in range(1000):
    if m.touch_status() == 1: # 'Y' touch pad
        pinY1.high()
        pyb.delay(1000)
        pinY1.low()

    if m.touch_status() == 2: # 'X' touch pad
        pinY2.high()
        pyb.delay(1000)
        pinY2.low()

    if m.touch_status() == 4: # 'B' touch pad
        pinY3.high()
        pyb.delay(1000)
        pinY3.low()

    if m.touch_status() == 8: # 'A' touch pad
        pinY4.high()
        pyb.delay(1000)
        pinY4.low()

    pyb.delay(200)
```

I found I could easily control the LEDs after the program was loaded and the board was reset. The MPR121 driver software made the pad detection quite easy to implement, and it is also very easy to modify the control program to suit additional requirements.

I will now transition to another program, which uses both the LCD and the touch sensors on the LCD module.

LCD and Touch-Sensor Demonstration

This demonstration comes from the www.micropython.org site, and it cleverly shows how to use the LCD to display changes as the touch pad sensors are activated. The following program is named lcddemo.py. It must be loaded into the same PYBFLASH USB drive as the mpr121.file. You do not load it into the main.py because it is run from the REPL prompt. Enter the following commands to run the program:

```
import lcddemo
lcddemo.run(1000)
```

The program listing follows with some extra comments that I have inserted to help clarify the program's operations.

```
import pyb
import mpr121

def run(n):
    lcd = pyb.LCD('X')
    lcd.light(True)
    m = mpr121.MPR121(pyb.I2C(1, pyb.I2C.MASTER))

    # this function creates the displayed block
    def blob(x, y, w, h, fill):
        for i in range(w):
            for j in range(h):
                if pyb.rng() & 0xff < fill: # pyb.rng()=30b random no.
                    lcd.pixel(x + i, y + j, 1)

    for i in range(n):
        t = m.touch_status() # any touch pad activated?
        lcd.fill(0) # fill the hidden buffer with 0's
        for y in range(32): # creates vertical divider
            lcd.pixel(64, y, 1)
        for x in range(128): # creates horizontal divider
            lcd.pixel(x, 16, 1)
        if t & 1: # 'Y' pad
            # 316-m.elec_voltage(0) creates the 'sprinkle' fill effect
            blob(90, 20, 10, 10, 316 - m.elec_voltage(0))
        if t & 2: # 'X' pad
            blob(30, 20, 10, 10, 316 - m.elec_voltage(1))
        if t & 4: # 'B' pad
            blob(90, 5, 10, 10, 316 - m.elec_voltage(2))
        if t & 8: # 'A' pad
            blob(30, 5, 10, 10, 316 - m.elec_voltage(3))
        lcd.show() # display the hidden buffer
        pyb.delay(50) # wait 50 ms
```

Figure 5-15 shows the LCD after the two commands have been entered at the REPL prompt. No pads have been touched at this point.

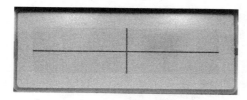

Figure 5-15 *Initial LCD screen.*

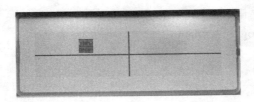

Figure 5-16 *The A touch pad activated.*

When I touched the A pad, a filled block appeared in the upper-left quadrant as shown in Figure 5-16.

This block stays on the screen as long as the A pad is activated or touched and the program is still running.

I then touched all the pads using my thumb and saw the resulting LCD as shown in Figure 5-17.

This last demonstration concludes the LCD and touch pad program discussion. I will now turn to the LCD module GPIO pin usage that I mentioned earlier.

Using the LCD Module GPIO Pins

Eight of the twelve touch-sensor inputs may also be used as GPIO pins, as introduced in a previous section. The first four sensor inputs are committed to the built-in touch pads and cannot be used as GPIO pins because the MPR121 chip firmware automatically configures these as touch inputs. In any case, eight GPIO inputs should be plenty to experiment with for our purposes.

The first thing that is needed prior to starting any demonstration is to solder a 14-pin female socket header to J3's 14 solder pads located immediately below the LCD on the module. This will allow an easy way to make connections to the GPIO pins, as well as to 3.3VDC and ground. Figure 5-18 shows the header soldered in place. Just be sure to keep the header as perpendicular to the board as possible.

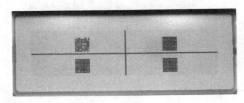

Figure 5-17 *All pads touched simultaneously.*

Figure 5-18   *Fourteen-pin header soldered to the LCD module.*

The mpr121.py driver software must now be modified to reconfigure the MPR121 chip so that touch inputs ELE4 to ELE11 are treated as GPIO pins instead of touch-sensor inputs. The key to this reconfiguration is to load a series of registers within the MPR121 chip with appropriate values. My initial reconfiguration simply made all the GPIO pins act as CMOS outputs set at a high level. I renamed the modified driver mpr121gpio.py in recognition of the added GPIO functions. I have provided a complete listing of the file here, as well as made it available for download from this book's companion website, www.mhprofessional.com/micropython.

```
"""
Modified driver for the MPR121 capacitive touch sensor.
This chip is on the LCD32MKv1 skin.
Incorporates GPIO functions
"""
import pyb
# register definitions
TOUCH_STATUS = const(0x00)
ELE0_FILT_DATA = const(0x04)
ELE0_TOUCH_THRESH = const(0x41)
DEBOUNCE = const(0x5b)
ELEC_CONFIG = const(0x5e)
GPIO_ENABLE = const(0x77)
GPIO_DIR = const(0x76)
GPIO_DATA = const(0x75)
GPIO_CONTROL0 = const(0x73)
GPIO_CONTROL1 = const(0x74)
```

```python
class MPR121:
    def __init__(self, i2c):
        self.i2c = i2c
        self.addr = 90 # I2C address of the MPR121

        # enable ELE0 - ELE3
        self.enable_elec(4)

    def enable_elec(self, n):
        """Enable the first n electrodes."""
        self.i2c.mem_write(n & 0xf, self.addr, ELEC_CONFIG)

    def threshold(self, elec, touch, release):
        """
        Set touch/release threshold for an electrode.
        Eg threshold(0, 12, 6).
        """
        buf = bytearray((touch, release))
        self.i2c.mem_write(buf, self.addr, ELE0_TOUCH_THRESH + 2 * elec)

    def debounce(self, touch, release):
        """
        Set touch/release debounce count for all electrodes.
        Eg debounce(3, 3).
        """
        self.i2c.mem_write((touch & 7) | (release & 7) << 4, self.addr, DEBOUNCE)

    def touch_status(self, elec=None):
        """Get the touch status of an electrode (or all electrodes)."""
        status = self.i2c.mem_read(2, self.addr, TOUCH_STATUS)
        status = status[0] | status[1] << 8
        if elec is None:
            return status
        else:
            return status & (1 << elec) != 0

    def elec_voltage(self, elec):
        """Get the voltage on an electrode."""
        data = self.i2c.mem_read(2, self.addr, ELE0_FILT_DATA + 2 * elec)
        return data[0] | data[1] << 8

    def enable_gpio(self):
        """Enable all the GPIO pins."""
        self.i2c.mem_write(0xff, self.addr, GPIO_ENABLE)

    def dir_gpio(self, n):
        """Set the pins for input or output (n is hex, 0 = in, 1 = out)."""
        self.i2c.mem_write(n, self.addr, GPIO_DIR)

    def data_gpio(self, n):
        """Set the data for the output pins (must match dir)."""
        self.i2c.mem_write(n, self.addr, GPIO_DATA)
```

```
def control0_gpio(self):
    """Setup control0 register"""
    self.i2c.mem_write(0x00, self.addr, GPIO_CONTROL0)

def control1_gpio(self):
    """Setup control1 register"""
    self.i2c.mem_write(0x00, self.addr, GPIO_CONTROL1)
```

The lcddemo.py program will also need to be modified somewhat to test the new GPIO functions. The renamed file is lcddemogpio.py and is listed next. It also is available from this book's companion website, www.mhprofessional.com/micropython.

```
import pyb
import mpr121gpio

def run(n):
    lcd = pyb.LCD('X')
    lcd.light(1)
    m = mpr121gpio.MPR121(pyb.I2C(1, pyb.I2C.MASTER))

    def blob(x, y, w, h, fill):
        for i in range(w):
            for j in range(h):
                if pyb.rng() & 0xff < fill:
                    lcd.pixel(x + i, y + j, 1)

    m.control0_gpio()
    m.control1_gpio()
    m.enable_gpio()
    m.dir_gpio(0xff)
    m.data_gpio(0xff) # Change this hex value for desired GPIO output


    for i in range(n):
        t = m.touch_status()
        lcd.fill(0)
        for y in range(32):
            lcd.pixel(64, y, 1)
        for x in range(128):
            lcd.pixel(x, 16, 1)
        if t & 1:
            blob(90, 20, 10, 10, 316 - m.elec_voltage(0))
        if t & 2:
            blob(30, 20, 10, 10, 316 - m.elec_voltage(1))
        if t & 4:
            blob(90, 5, 10, 10, 316 - m.elec_voltage(2))
        if t & 8:
            blob(30, 5, 10, 10, 316 - m.elec_voltage(3))
        lcd.show()
        pyb.delay(50)
```

The new lcddemogpio.py program is run in an identical fashion to the way the unmodified lcddemo.py was run. Simply enter these commands at the REPL prompt:

```
import lcddemogpio
lcddemogpio.run(1000)
```

You will still be able to use the touch pads and have the blocks appear on the LCD, but the GPIO pins ELE4 to ELE11 will now be all active highs. Incidentally, the actual pins on the header starting from the left side are ELE11 to ELE0, with the last two being 3V3 and GND, respectively. You should refer to the schematic in Figure 5-10 to refresh yourself on the module pin-out.

Please note that the GPIO pin levels are not interactive, meaning that the levels are fixed in the Python code and not set from a user prompt. My purpose in this section was to simply show how to make some additional GPIO pins available using the MPR121 chip. Readers are encouraged to modify this code to accommodate their requirements, including interactivity if so desired.

CAUTION   *I do wish to point out that there is a 12-mA output current limit for each of the GPIO pins. This means that you should use a 330-Ω current-limiting resistor if you desire to drive an LED from any of the GPIO pins.*

The next section discusses some additional MPR121 GPIO features that may interest some readers, especially those interested in experimenting with pulse width modulation (PWM).

MPR121 PWM

The material in this section is based upon the Freescale application note AN3894, titled "MPR121 GPIO and LED Driver Function." The application note content is not duplicated in the MPR121 datasheet, which is one major reason that I like to review manufacturer's notes as well as their general overview datasheets. I discovered in the note that each of the GPIO pins is capable of being pulse width modulated, which lends them to adding control functions to the MPR121 chip such as intensity control, that is, dimming, and even servo control to some extent. PWM will be enabled based upon the content of the PWM registers. It is a bit odd that these registers are not even discussed in the general MPR121 datasheet. Table 5-3 details the PWM register addresses, as well as how the register content is arranged to control individual GPIO pins.

Name	Address	D7	D6	D5	D4	D3	D2	D1	D0
PWM 0	0x81	PWM1[3]	PWM1[2]	PWM1[1]	PWM1[0]	PWM0[3]	PWM0[2]	PWM0[1]	PWM0[0]
PWM 1	0x82	PWM3[3]	PWM3[2]	PWM3[1]	PWM3[0]	PWM2[3]	PWM2[2]	PWM2[1]	PWM2[0]
PWM 2	0x83	PWM5[3]	PWM5[2]	PWM5[1]	PWM5[0]	PWM4[3]	PWM4[2]	PWM4[1]	PWM4[0]
PWM 3	0x84	PWM7[3]	PWM7[2]	PWM7[1]	PWM7[0]	PWM6[3]	PWM6[2]	PWM6[1]	PWM6[0]

Table 5-3 *PWM Registers*

There is also one other set of references you need to be aware of, and that is the correspondence between the ELECTRODE (ELE), GPIO, and PWM designations. Table 5-4 details the correspondence between the designations given that PWMx = GPIOx for Tables 5-3 and 5-4.

Overall, the PWM0[3:0] to PWM7[3:0] register bits are used to set the PWM duty of GPIO0 to GPIO7, respectively. Upon a power-up reset, the default setting for each of these four PWM registers is 0x00. This value means that PWM is not in effect and the corresponding data register value is continually output. A GPIO pin will output a PWM waveform when it is programmed as output and the corresponding data register bit for that pin is high or 1 and the corresponding PWMx[3:0] register is nonzero. The PWM period is fixed at 8 ms, which is 1/256 of the internal MPR121 32-kHz oscillator frequency. The PWMx [3:0] register fixes the duty cycle of the PWM waveform. The on time for a PWM pulse ranges from 0.5 to 7.5 ms because each PWM register uses four bits to determine its value. You get a high-level pulse interval of 0.5 ms when the 8-ms period is divided by 16, the maximum interval number produced using four bits.

There are likely to be inaccuracies in the PWM duty cycle because GPIO output transition is inhibited during a touch electrode measurement state. Thus, when PWM interval time is close to electrode measurement time, the PWM operation will be disturbed and there will be a disturbance in the PWM waveform. This situation is unavoidable because PWM operations are secondary to electrode operations, as that is the primary purpose for the MPR121 chip.

I further modified both the mpr121gpio.py and lcddemogpio.py files to demonstrate a PWM waveform on GPIO0, which according to Table 5-5 is also ELE4. I did not provide full lists for either one of these files, but put just the snippets of the additional code that needed to be added to each one. I renamed the complete files mpr121pwm.py and lcdpwm.py, and they are available from this book's

PIN #	8	9	10	11	12	13	14	15	16	17	18	19
ELECTRODE	ELE0	ELE1	ELE2	ELE3	ELE4	ELE5	ELE6	ELE7	ELE8	ELE9	ELE10	ELE11
GPIO	—	—	—	—	GPIO0	GPIO1	GPIO2	GPIO3	GPIO4	GPIO5	GPIO6	GPIO7

Table 5-4 *Correspondence between ELE and GPIO Designations*

companion website, www.mhprofessional.com/micropython. The code snippets are listed here:

```
# Added to mpr121gpio.py
GPIO_PWM0 = const(0x81)
def pwm0_gpio(self, n):
        """Sets the PWM 4-bit dividers for GPIO0 and GPIO1."""
        self.i2c.mem_write(n, self.addr, GPIO_PWM0)
# Added to lcddemogpio.py
m.pwm0_gpio(0x03)
```

Figure 5-19 is a screen capture of the GPIO0 (ELE4) PWM waveform taken by my USB oscilloscope.

As you readily see, the high portion of the waveform lasts for 1.5 ms, which was fixed by the 0x03 value sent by the m.pwm0_gpio(0x03) method call.

This last waveform should be of some interest for those folks interested in controlling a standard servo. The 1.5-ms PWM waveform will direct a servo to remain in its neutral position. Sending a servo a 1.0-ms pulse waveform will cause it to travel to its extreme counterclockwise (CCW) position. Similarly, sending a 2.0-ms pulse waveform will cause it to rotate to its extreme clockwise (CW) position. Now admittedly, controlling a servo to only three positions is likely to be of limited value, but there may be situations where this feature can come in handy. The other point to consider is that the MPR121 chip can be configured for up to eight servo channels, which is an intriguing point for some cost-sensitive designs.

This last section concludes this chapter on the LCD module.

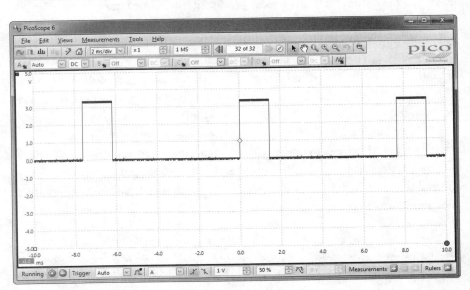

Figure 5-19 *GPIO0 PWM waveform.*

Summary

I started this chapter with an overview of the LCD and touch-sensor skin (board), which I purchased from the MicroPython store. This relatively inexpensive board or module comes with some rather impressive capabilities, especially regarding the touch sensors, which are controlled by the MPR121, a second-generation Freescale chip.

Some fairly standard monochrome LCD functions were first demonstrated, including character and graphical displays. I also discussed how the LCD class found in the pyb library makes programming the module's display very simple to accomplish. In addition, I showed you how to use an external command with the display. This command is part of the Sitronix LCD controller firmware. The LCD class method named command was used to access the firmware command.

The concepts behind capacitive sensing were next presented, which laid the framework for a detailed discussion about the Freescale MPR121 capacitive touch-sensor chip. I showed you some of the key registers that must be programmed to properly use this chip.

A detailed discussion of the MPR121 driver software followed, which concluded with an initial demonstration of the four built-in touch pad sensors.

Next followed a simple experiment where each touch pad controlled a dedicated LED. This test demonstrated how to control a Pyboard GPIO pin with a touch pad sensor input. An LCD and touch pad sensor demonstration was shown based upon a great program made available from the www.micropython.org website.

The next several sections discussed how eight sensor inputs could also be used as GPIO pins. I showed you how to modify the existing mpr121.py driver program to incorporate the GPIO functionality.

The chapter concluded with a brief introduction on how the MPR121 chip also supports PWM on its eight GPIO pins. I mentioned that the PWM function could even be made to control standard servos, albeit in a limited fashion.

6

The AMP Audio Skin

In this chapter I will show you how to install and program an AMP audio skin (board) that I purchased from the MicroPython store. I will explore the audio input and output functions of this board, which are completely supported by the MicroPython language.

Assembling the AMP Board

The backside of the AMP board is shown in Figure 6-1 and comes with three male headers, two 8-pin and one 3-pin. These need to be soldered onto the board to allow the AMP board to be plugged into the Pyboard. I would suggest using a solderless breadboard to hold the male pin headers in place while you solder them to the board.

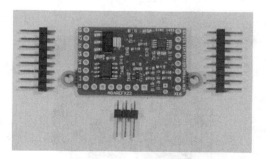

Figure 6-1 *Backside of the AMP board.*

Figure 6-2 *Front side of the AMP board.*

The front side of the AMP board is shown in Figure 6-2, where you may clearly see all the labeled solder pads for the male header pins. Note that the three-pin header is perpendicular to the other eight-pin headers because it is needed to connect the AMP board to one of the Pyboard's analog-to-digital converters (ADCs), along with the analog voltage supply and ground connections.

Note that the speaker's two wires must be soldered to the labeled speaker solder pads located near the lower-right corner of the board. Although technically not required, I would suggest soldering the red speaker lead to the square solder pad and the black lead to the remaining round pad.

Additionally, the electret microphone must be soldered to the pads located at the left of center on the bottom row of solder pads. You must carefully place this tiny microphone into the two pads such the microphone case just covers the circular silkscreen image on the board. This is necessary because the microphone leads are polarized, meaning there are plus and minus leads, which must match the board connections. Figure 6-3 is a macro-photograph of the underside of the electret microphone. You should be able to see the + symbol next to the upper pin. This must be inserted into the matching + pin on the AMP board.

Figure 6-3 *Macro-photograph of the electret microphone.*

Figure 6-4 *Front side of fully assembled AMP board.*

Figure 6-4 shows the front side of the fully assembled AMP board with both the speaker and microphone mounted.

AMP Board Circuits

The AMP audio board is a compact module that is designed to play audio generated by the Pyboard's digital-to-analog converter (DAC). The AMP board also contains a digital potentiometer, which controls the gain or volume of the audio signal being input to the onboard power audio amplifier. Figure 6-5 shows a portion of the AMP board schematic that comprises the audio output signal chain.

The AMP board also contains an audio input section, which uses an electret microphone to detect speech and other audible sounds, which are then passed to a pre-amplifier and then into one of the Pyboard's ADCs. Figure 6-6 shows the portion of the AMP board schematic that comprises the audio input signal chain.

You should note that the schematic shows a pair of solder pads brought out to a connection labeled P2. It appears that the connection probably was used for

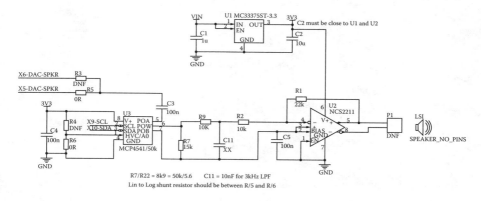

Figure 6-5 *Audio output signal chain schematic.*

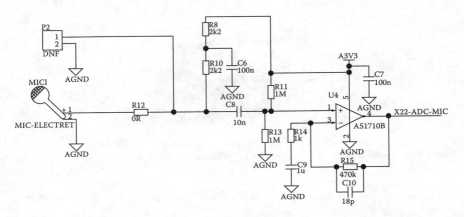

Figure 6-6 *Audio input signal chain schematic.*

debugging and testing during the design process, as there is no actual P2 connector mounted on the board. I just wanted to mention it as a possible injection point for low-level signals that need some additional gain prior to being input into the Pyboard's ADC. You would also need to unsolder the 0-Ω SMT R12 resistor that connects the electret microphone to the circuit so as to prevent any interference with the applied external signal.

Figure 6-7 shows all of the pin connections between the AMP module and Pyboard.

It should be reasonably apparent that there is only one way to plug in the AMP module into the Pyboard. In any case, Figure 6-8 shows the AMP module mounted on the Pyboard.

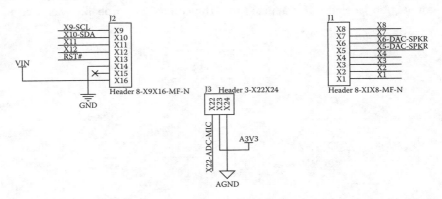

Figure 6-7 *Pin connections between the AMP module and Pyboard.*

Figure 6-8   *AMP module mounted on the Pyboard.*

I will next separately discuss each of the main components for both the audio input and output circuits to provide you with a good background and perhaps trigger some ideas on where you might use similar components in other projects.

Audio Input Circuit

There are only two main components in the audio input circuits. These are the electret microphone and the pre-amplifier chip.

Electret Microphone

The electret microphone mounted on the AMP module is extremely inexpensive, yet has impressive specifications, including a 20-Hz to 20-kHz frequency response. Figure 6-9 shows a typical electret microphone.

Figure 6-9   *Typical electret microphone.*

Figure 6-10 *Inner components of the electret microphone.*

This figure was excerpted from the website http://www.openmusiclabs.com/ learning/sensors/electret-microphones/ from which much of the remaining material in this section is based.

The inner components constituting the microphone are shown in Figure 6-10.

The essential operation of the microphone is fairly simple and can be explained by the equivalent electrical schematic shown in Figure 6-11.

The circuit element labeled C in the diagram is the actual microphone transducer. It is made up of an extremely thin circular piece of metalized Mylar film mounted on a washer, which holds a fixed electrical charge. This Mylar sheet is positioned over, but not quite touching, a metal cover plate holding very high-impedance junction field effect transistor (JFET). The gate of the JFET is connected to the metal cover. This arrangement forms a capacitor where vibrations from impinging acoustic waves will cause slight deformations in the Mylar sheet. These deformations will cause the charge on the sheet to vary in distance from the cover plate, inducing a voltage potential to be created at the JFET gate. There is almost negligible current flow in the JFET because it has extremely high input impedance.

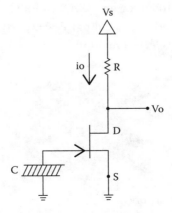

Figure 6-11 *Equivalent electrical schematic for an electret microphone.*

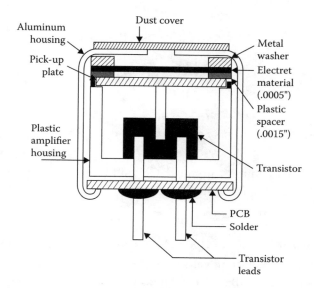

Figure 6-12   *Cut-away diagram for an electret microphone.*

The varying voltage present on the two JFET output leads, which are also the microphone leads, represents the electrical analog of the sound waves entering the small hole in the external case, shown in Figure 6-10.

Finally, Figure 6-12 is a cut-away diagram showing how all the microphone parts are arranged in this clever device.

Pre-amplifier

A pre-amplifier is required in this input circuit because the voltage levels generated by the electret microphone are very low. Consequently, the Pyboard's ADC could not produce any meaningful output without a significant increase in the signal level input. The pre-amplifier used is an Austria Mikro Systeme, model AS17170 single supply, low-offset operational amplifier (or simply op-amp). This op-amp can operate using a supply voltage (Vcc) ranging from 2.7 to 5.5 VDC. In this circuit, it is operated with 3.3 VDC provided by the Pyboard. The op-amp is also classified as a rail-to-rail device, meaning that it is capable of generating a peak output voltage very close to the rail or Vcc supply voltage.

The op-amp is configured as a noninverting amplifier, which means the pre-amp's output is in phase with the input signal. Figure 6-13 is a simplified schematic for a noninverting op-amp.

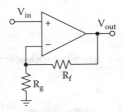

Figure 6-13 *Generic noninverting op-amp circuit.*

The overall static (direct current [DC]) gain from this circuit may easily be calculated using the following equation:

$$Av = 1 + Rf/Rg$$

In the actual circuit, $Rf = 470K$ and $Rg = 1K$. Therefore, $Av = 471$. Amplifier gains are typically specified in voltage decibels (dB), so a fixed gain of 471 would be equivalent to a 53.46 dB gain, computed as shown:

$$A = 20 \log(471) = 53.46$$

This is a significant amount of gain, which should be sufficient to easily amplify the low-level electret microphone signals. One more item that you should be aware of is the op-amp's gain-bandwidth specification. This parameter places an upper limit on how much gain an op-amp can produce. It turns out that the maximum gain available is frequency dependent, lowering as the frequency increases. Figure 6-14 shows the gain-bandwidth curve for this op-amp excerpted from the device's datasheet.

You can see that I placed two dashed lines on the curve showing the maximum possible gain available at 20 kHz is approximately 54 dB. This value nicely aligns with the static gain already determined by the circuit resistor values I discussed earlier.

The last item I wanted to mention about the op-amp is that it is a relatively low-noise device. It is important that a pre-amplifier generate very low internal noise because high gain values will greatly increase this noise to the point of it seriously interfering with overall pre-amplifier operation.

I will defer presenting a demonstration of how the audio input circuit's function until after the audio output circuit is discussed, as the input circuit demonstration requires the output circuit.

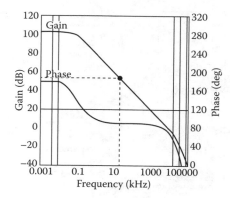

Figure 6-14   *Gain-bandwidth curve.*

Audio Output Circuit

There are three main components in the audio output circuits. These are the digital potentiometer, the audio power amplifier, and the low drop-off voltage regulator chip.

The Digital Potentiometer

The AMP board digital potentiometer chip is the model MCP4541 manufactured by the Microchip Corporation. The purpose of the digital potentiometer is to act as a digitally controlled resistive network, which controls the amplitude of the audio signal sent to the audio power amplifier. Figure 6-15 shows the MCP4541 block diagram that includes the resistive network used in the chip. This figure was excerpted from the MCP4541 datasheet.

The resistive bridge between the P0A and P0B terminals is made up of 128 individual resistors, which are connected in series to form a 50-KΩ resistance. The P0W (wiper) terminal is digitally controlled to connect to any of the 127 connection points between the individual resistors. The inter-integrated circuit (I2C) serial-bit protocol is used to interface the digital potentiometer to the Pyboard. The digital potentiometer's I2C address is set at 0x2e, or decimal 46. The I2C address can be set to 0x2f by setting the A0 pin to high. Referring to Figure 6-5, if the R6 0-Ω SMT resistor is removed and soldered into the R4 position, this will cause the A0 pin to be set at a high level, which changes the I2C address to 0x2f.

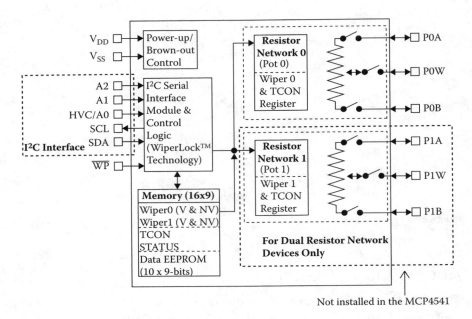

Not installed in the MCP4541

Figure 6-15 *MCP4541 block diagram.*

The following function is used to control the signal volume sent to the audio power amplifier.

```
def volume(val):
    pyb.I2C(1, pyb.I2C.MASTER).mem_write(val, 46, 0)
```

The val variable may range from 0 to 127, where 127 is the maximum volume setting. I will discuss the volume setting in more detail in the initial test section.

Audio Power Amplifier

The audio power amplifier is the model NCS2211 manufactured by ON Semiconductor. It is formally called a "Low Distortion Audio Power Amplifier with Differential Output and Shutdown Mode." The schematic in Figure 6-16 shows the chip connected to a speaker in the audio output circuit.

The differential output is also called class A/B and is very efficient in its operation. This amplifier achieves a full 1-W output into a "bridged" 8-Ω speaker with a very low distortion of only 0.2 percent third harmonic distortion (THD).

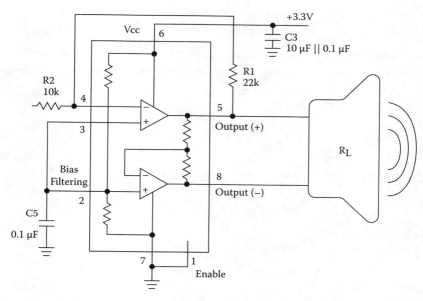

Figure 6-16 *Audio power amplifier circuit.*

Low Drop-Off Voltage Regulator

The third and final main component in the audio output circuit is the model MC33375ST-3.3 manufactured by ON Semiconductor. It is formally called a "300 mA, Low Dropout Voltage Regulator with On/Off Control." This regulator uses the 5-VDC VIN voltage from the Pyboard and converts it to 3.3 VDC for use by the audio power amplifier. This chip supplies about 300 mA at 3.3 VDC when the audio amplifier is producing the full 1-W output.

One more item to be mindful of is to ensure that the USB port provides 500 mA, if that is the way you are powering the AMP module/Pyboard combination. Some USB ports only provide 100 mA, which is simply not sufficient to allow the regulator to drive the audio power amplifier to its maximum volume and will likely result in poor-to-ineffective amplifier operation.

It is now time to show you how to test the AMP module.

Initial Test

You should enter the following code into the PYBFLASH USB drive to start the initial tests. I named the file demoAMP.py and it is available from this book's companion website, www.mhprofessional.com/micropython.

```python
import pyb
import math
from pyb import DAC

def volume(val):
    pyb.I2C(1, pyb.I2C.MASTER).mem_write(val, 46, 0)

def run():
    # create a buffer containing a sine-wave
    buf = bytearray(100)
    for i in range(len(buf)):
        buf[i] = 128 + int(127 * math.sin(2 * math.pi * i / len(buf)))

    # output the sine-wave at 440Hz, middle-C
    dac = DAC(1)
    dac.write_timed(buf, 440 * len(buf), mode=DAC.CIRCULAR)
```

Next open a read-eval-print-loop (REPL) session and enter the following commands:

```python
import demoAMP
demoAMP.volume(127)
demoAMP.run()
```

If all goes well, you should now be hearing a 440-Hz tone coming from the AMP speaker. The audible tone I heard was not too loud for reasons I cannot explain. It maybe that the USB port I was using was a bit underpowered or the speaker is simply too small to create more sound intensity. In any case, the AMP module was working properly; however, I detected some moderate distortion in the tone, which made me consider troubleshooting the source of the distortion. I have detailed the troubleshooting steps I used in the following section. You may choose to read it if it is of interest or simply skip it without losing any continuity in the overall chapter content.

Troubleshooting the Distortion

As an initial troubleshooting step, I always want to capture more information about the problem. I have a USB oscilloscope (scope)—an invaluable tool in gaining additional information about circuit and system problems. I have used the scope in previous chapters to display waveforms. In this situation I used the scope

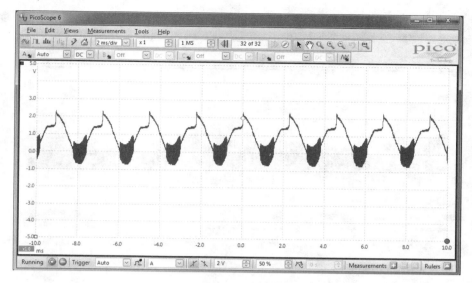

Figure 6-17 *440-Hz USB oscilloscope trace at maximum volume.*

to closely examine the output signal going to the speaker. I had to solder two additional wires to the speaker pads in order to have connection points for the scope. This is a bit tricky, as there is very little spacing between the pads and it is easy to create a solder bridge, which would be disastrous for the audio power amplifier. However, I was successful, and Figure 6-17 shows the time-based waveform for this 440-Hz tone.

As you can clearly see from the figure, the actual waveform bears little resemblance to a pure 440-Hz sine wave other than the fact that the period is precisely the same at 2.27 ms. This waveform is an obviously highly distorted and contains many harmonics, which I confirmed by switching the scope to the spectrum or frequency display. Figure 6-18 shows the equivalent spectrum for the time trace shown in Figure 6-17.

Strong harmonic spikes are clearly evident at 880 Hz, 1320 Hz, 1760 Hz, and so on, which all indicate the presence of moderate-to-severe distortion in the baseband tone. I added a total harmonic distortion measurement to the scope, which showed that a level slightly over 20 percent was present in the signal.

I then reasoned that distortion might be due to overdriving the audio power amplifier, so I cut the volume in half and repeated the same test. Figure 6-19 shows the time-based trace at this volume.

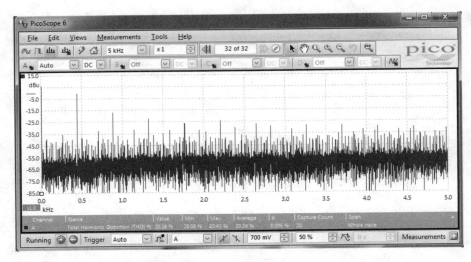

Figure 6-18 *Frequency spectrum for the 440-Hz tone.*

The trace still clearly shows significant distortion, which rules out the volume dependency theory. The next troubleshooting step was to examine the input signal going to the AMP module. This signal originates from pin X22, which is one of the Pyboard's DAC outputs. I simply held the scope probe to the X22 solder pad and captured the waveform shown in Figure 6-20.

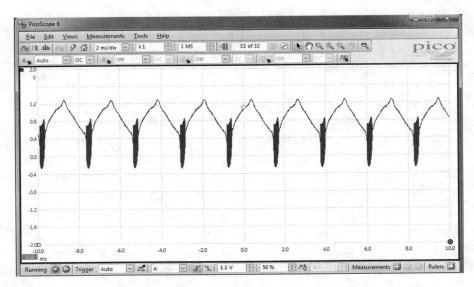

Figure 6-19 *440-Hz USB oscilloscope time trace at reduced volume.*

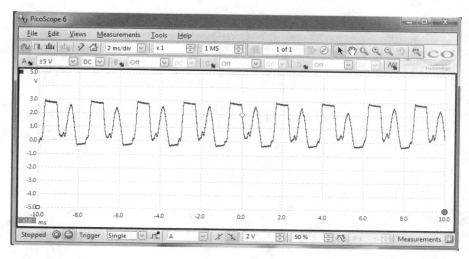

Figure 6-20 *DAC output.*

It is clearly distorted, but not as badly as the signal output from the audio power amplifier. The scope trace also shows the presence of clipping, which is indicative of the DAC being overdriven. This result quickly led me to the next step, which was to change the sine wave generation software to produce a waveform with less amplitude and, hopefully, little to no distortion. I made the following change to the demoAMP.py program, which effectively cut the DAC output in half:

```
buf[i] = 64 + int(63 * math.sin(2 * math.pi * i / len(buf)))
```

Figure 6-21 shows the new DAC output resulting from this software modification.

Although not perfect, it certainly is an improvement over the waveform shown in Figure 6-20.

The next step in the troubleshooting sequence was to go back to the audio power amplifier output and capture a new trace to see if the distortion problem was fixed. Figure 6-22 shows a fairly clean 440-Hz sine wave, which indicated to me that the distortion was indeed sharply reduced to what I would consider an acceptable level.

I also captured a frequency spectrum of this sine wave in Figure 6-23 to determine the actual levels of any remaining harmonic frequency spikes contributing to the distortion.

If you closely examine the figure you will see that the second harmonic at 880 Hz is approximately −40 dB below the fundamental 440-Hz level.

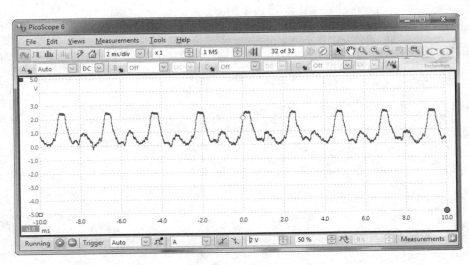

Figure 6-21 *DAC output with reduced amplitude.*

The amplitudes of the remaining higher-order harmonics are approximately the same or lower than the second harmonic. A 40-dB reduction is the same as being 1/100th of the fundamental amplitude. There is approximately 2 to 3 percent THD in the final signal. It would be very hard for the average listener to detect any audible distortion in this 440-Hz, or middle-C, tone.

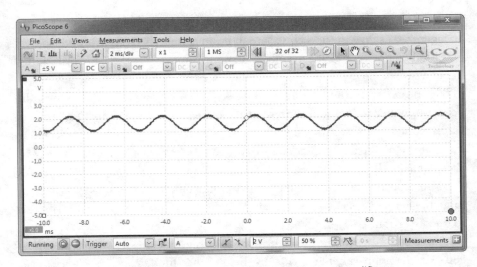

Figure 6-22 *440-Hz sine wave output from the audio power amplifier.*

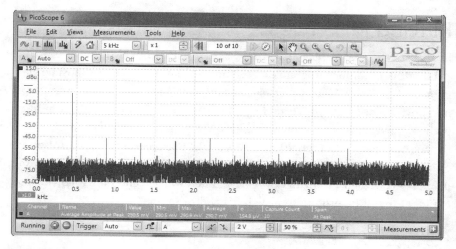

Figure 6-23 *Frequency spectrum of the 440-Hz sine wave.*

At this point, you might be asking yourself the following: What if I cleaned up the distortion that obviously was still present in the DAC output signal? The answer lies in the use of a single-pole, low-pass RC filter shown as a schematic in Figure 6-24. There is also a Bode frequency response chart shown in the figure.

This filter is installed at the input of the audio power amplifier for the obvious purpose of reducing higher-frequency artifacts, which would normally introduce some signal distortion.

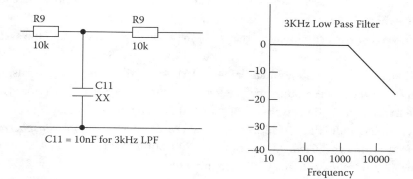

Figure 6-24 *3-kHz low-pass filter.*

Playing MP3 Files

It is relatively easy to play eight-bit .wav files using the AMP module with the Pyboard. Two support programs are required to play .wav files that must be placed in the PYBFLASH USB drive. These programs are wave.py, available from https://micropython.org/static/doc/examples/wave.py, and chunk.py, available at micropython.org/resources/examples/chunk.py. You will also need an eight-bit .wav file to play. An example file named test.wav is available at MicroPython.org/resources/examples.test.wav. After downloading, store the two support files and the example .wav file in the PYBFLASH USB drive. Then you can enter the following commands at the REPL prompt to play the .wav file example:

```
import wave
from pyb import DAC
dac = DAC(1)
f = wave.open('test.wav')
dac.write_timed(f.readframes(f.getnframes()),
f.getframerate())
```

The .wav file should immediately be heard playing from the AMP board speaker. This .wav file is just a short sound burst consisting of three string notes played in rapid sequence. The test.wav file is only 9 kB, and it lasts for about one second. You should also be aware that the .wav file to be played is loaded into the heap by the chunk.py program. The heap memory consists of the 100-kB RAM storage on the Pyboard. There really isn't much room left for a .wav file after the wave.py and chunk.py files are loaded into the heap. Attempting to play a .wav file whose duration exceeds two seconds will likely result in a memory allocation error, which I constantly endured while trying to play a longer .wav file. All in all, the AMP/Pyboard combination really doesn't provide much of a useable system to play .wav files unless it just a short beep or squeak. So until the Pyboard is modified to incorporate a much larger RAM, I would suggest avoiding using it along with the AMP module for playing .wav files. Many other small boards are available, including the Raspberry Pi, which are ideally suited for playing .wav and other similar audio files. In addition, I would recommend my book, *Raspberry Pi Projects for the Evil Genius* (McGraw-Hill Education, 2014), which has a project showing how to use it to play .wav and .mp3 files.

It is now time to move on to demonstrate how the audio input section functions, having shown you how the audio output works.

Audio Input Demonstration

The electret microphone is the sensor used for audio inputs. It has a relative broad frequency response, as previously discussed. I used an app on my iPhone to generate a 500-Hz audio tone, which the microphone received and sent on to the Pyboard's ADC. I used the following commands at the REPL prompt to both record the digitized audio tone and play it back through the AMP board speaker:

```
# make a digital recording
dac = pyb.DAC(1)           # create a DAC object
buf = bytearray(20000)     # create a 20000 byte array to store samples
adc = pyb.ADC('X22')       # setup the X22 pin ADC channel
adc.read_timed(buf, 20000) # start the ADC sampling
20000                              # 20000 appears when sampling ends

# playback the digital recording
dac.write_timed(buf, 20000) # 1 second duration 500Hz tone
```

The 500-Hz tone played very clearly, although at a highly reduced volume for reasons that I discuss later in the section.

It is also easy to change the playback frequency. Simply enter the following command to lower the tone to 125 Hz. Note that the note will now be four seconds in duration because the playback frequency is one quarter of the original 20,000 Hz.

```
dac.write_timed(buf, 5000)
```

Likewise, it is just as easy to speed up the playback by increasing the frequency argument. The following command will change the recorded tone from 500 Hz to 750 Hz. Consequently, the tone duration will shorten to 0.75 seconds.

```
dac.write_timed(buf, 30000)
```

Use this next command to continuously loop the tone playback. I did this in order to measure both the time trace and spectrum of the recorded 500-Hz tone.

```
dac.write_timed(buf, 20000, mode=pyb.DAC.CIRCULAR)
```

Figure 6-25 shows the 500-Hz looped tone time trace as captured from the AMP module speaker output.

The signal waveform looks very "clean" with little distortion present. I confirmed the THD with a spectrum capture shown in Figure 6-26.

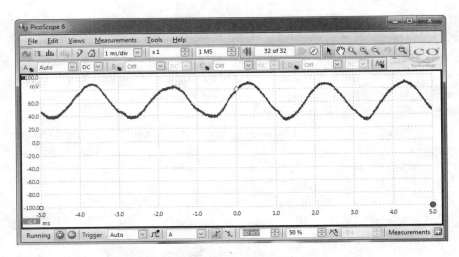

Figure 6-25 *Time trace for 500-Hz continuous looped test tone.*

The total harmonic distortion was measured at about 4.5 percent, which is still fairly low and would not be too objectionable in most audio signal replays. I would, however, like to comment on the reduced signal amplitude, which is shown in Figure 6-25. The peak signal amplitude shown on the trace is about 90 mV. This is quite a significant reduction from previously displayed signal

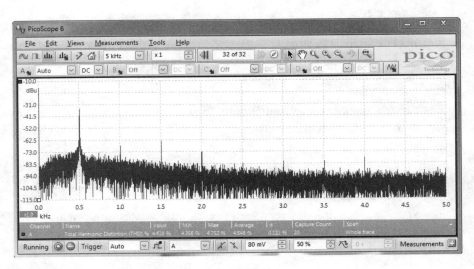

Figure 6-26 *Spectrum for the looped 500-Hz tone.*

amplitudes created by the demoAMP program. The reason this happens is that I used a bytearray buffer to store the ADC samples. Obviously, the bytearray only uses 8 bits to store a digital sample, but the Pyboard ADC produces a 12-bit sample. Consequently, the sample's lower four bits are discarded and the remaining eight bits are used instead to represent the whole sample. This means that the peak absolute voltage that can be stored is drastically reduced from 3.3 VDC to approximately 0.2 VDC, or a one-sixteenth reduction in the input voltage range. This can be avoided by using two bytes to store each sample; however, the resulting software to support two bytearray elements becomes much more complex and not too relevant for my demonstration purposes. I just point this out as a consequence of using a bytearray buffer so you will know why your replayed signal is much weaker than the original.

This last section concludes this chapter on how to install and use the AMP module with the Pyboard and MicroPython language.

Summary

The purpose of this chapter was to show you how to install and use the AMP skin module with a Pyboard. This module is available from the MicroPython store. A small amount assembly is necessary before the module can be used. I would advise you to have a fine-tipped, temperature-controlled soldering iron available, which would greatly help with the assembly.

After the assembly section, I discussed all of the main components mounted on the module. This discussion was provided a general background on these parts and pointed out several inherent limitations and constraints of the module due to the selection of these circuit components.

An initial test demonstration followed where a tone was generated by the Pyboard's DAC and subsequently played through the module's speaker. The tone was somewhat distorted, which led me to present a section where I went through a series of troubleshooting steps in order to successfully identify and remedy the source of this distortion.

I next presented a brief demonstration on how to play a .wav file using the Pyboard and AMP module. Playing .wav files is of limited use with a Pyboard due to the rather small amount of dynamic memory (RAM) present on the board. It is not really recommended that the Pyboard be used for this particular application.

The chapter concluded with a demonstration of the audio input circuit. I showed you how to record and play back an audio tone produced by an iPhone app. I also demonstrated how to change the pitch or frequency of the tone, raising or lowering it as desired. The section ended with a brief advisory on using bytearrays to store digital samples from a Pyboard ADC. An inherent 12- to 8-bit conversion will severely limit the peak output voltage, which results in a reduced volume from the speaker.

7

An Autonomous Robotic Car

In this chapter I will show you how to build and program a small, autonomous robotic car that is able to navigate a simple obstacle field without any outside intervention.

Building the Robot Car Platform

The basic robotic car is shown in Figure 7-1. It was constructed from a kit available from the Parallax Corporation, www.parallax.com.

Figure 7-1 *Basic robotic car.*

Figure 7-2 *Underside of the robotic car.*

Parallax named this model the Boe-Bot, which is short for Board of Education Robot. The company sells several versions of the Boe-Bot, which include different microcontrollers. I happened to purchase their kit controlled by their Propeller Activity Board. However, a model 28124 parts kit is available without a microcontroller, which will allow you to build only the robot car platform and then add your own Pyboard.

The car is powered by two continuous rotation (CR) servos, which may be seen in Figure 7-2, showing the underside of the robot.

It is easy to quickly build the basic robotic car platform. You can even go to the Parallax website and download the assembly instructions if you want a preview of what is involved in the platform assembly. However, I will first discuss how CR servos function before discussing how I built the robot car platform. It is fairly important to understand how these devices function, as they are prime movers for this project as well many other similar applications. The Pyboard must generate specific control signals in order to control these devices, which is why I am including the following sections on how both standard and CR analog servos function.

A Standard Analog Servo

Figure 7-3 shows a view of the internals of a standard analog servo commonly used in radio-control (RC) applications. I have labeled the main components in the figure, which are always found in an analog RC servo.

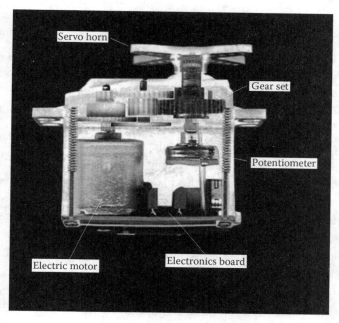

Figure 7-3 *Internal view of an analog servo.*

The electric motor shown in the figure is typically an inexpensive hobby-style electric motor, which normally rotates at a no-load rotational speed of approximately 12,000 rpm. The output torque of this motor type is very low, in the range of 0.1 oz-in, which is far too low to be of much use in a typical RC servo application. A speed reduction gear set is therefore used in the servo to increase the output torque to a useable level. The typical speed reduction ratio implemented by the gear set is approximately 400 to 1, which means the rotational speed goes from 12,000 to 30 rpm. However, the torque increases by precisely the same ratio, meaning the output torque is increased to 40 oz-in, which is more than adequate for most RC servo applications. The servo's much lower 30-rpm rotation speed is also quite compatible with RC applications.

A servo horn is shown attached to the top of the output gear shaft in the figure. This horn is simply a plastic piece that connects the output shaft to an RC actuator, which could be a model aircraft control surface, such as a flap or aileron. A potentiometer is also connected to the bottom of the output shaft. The purpose of the potentiometer is to create a feedback voltage, which is sent to the electronic control board for servo output shaft positioning. Figure 7-4 illustrates the relationship between servo shaft position and the input control signal generated by the Pyboard.

You should be able to see from the figure that a repetitive 1-ms-wide pulse will cause the servo to position the output shaft to a 90-degree position counterclockwise (CCW) from the neutral position. The repetition frequency is typically set at 50 Hz but it may be set higher at or slightly above 100 Hz. This type of repetitive pulse signal is known as pulse width modulation (PWM) because the width of the control pulse directly controls the servo operation.

A PWM signal with a 1.5-ms width will cause the servo shaft to be set at a neutral position. Finally, a 2.0-ms PWM pulse will cause the servo to position the output shaft at a 90-degree position clockwise (CW) from the neutral position. Of course, the shaft can be positioned between the CCW and CW limits by applying the appropriate PWM signal. For instance, a 1.75-ms PWM signal will cause the shaft to be set at 45 degrees CW from the neutral position.

The potentiometer attached to the output shaft is the means by which the physical shaft rotation is measured. The potentiometer is at its midway point when the servo is at its neutral position. A typical analog servo is powered by 6 VDC, which means 3 VDC is sent from the potentiometer to the electronics control board when the shaft is at the neutral position. Likewise, a voltage of approximately 1 VDC represents the full CCW limit, and 5 VDC represents the CW limit. The electronics control board utilizes analog control circuits such as multivibrators and voltage comparators to control the electric motor based upon

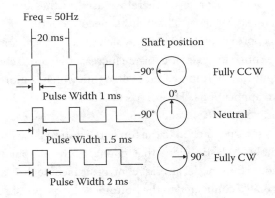

Figure 7-4 *Servo shaft position and control signal waveforms.*

the input PWM signal and the potentiometer feedback control signal. The control board is actually fairly simple and has been refined over many years to be very reliable and stable. The same control board design is used in most inexpensive analog RC servos. The CR servo is a simple modification of the standard servo and is discussed in the next section.

CR Servo

A standard servo is easily adapted to become a CR servo by removing the potentiometer and replacing it with a fixed resistor voltage divider that always generates the midpoint or neutral shaft position voltage. This means that when a 1.5-ms PWM is applied, the servo will be in its neutral position and the electronics control board will be sensing the neutral shaft position voltage so the electric motor will not be powered on, which is precisely what is desired. However, if a 1.0-ms PWM signal is applied, the control board will still be sensing the neutral voltage and will then power on the electric motor in a CCW direction, trying to reduce the feedback voltage so that it just balances the voltage it expects for the fully CCW position limit. Of course, this never happens because a potentiometer is no longer present and the motor continues to run at a full CCW rotational speed.

The opposite happens when a 2.0-ms PWM signal is applied to the CR servo. It will run at the full CW rotational speed. Just remember that the full rotational speed in either the CW or CCW directions is only 30 rpm due to the gear train.

It is also true that the CR servo can be set to intermediate speeds by applying appropriate PWM signals in a similar manner as how the standard servo could be set to intermediate positions by applying PWM signals between the neutral and full limit values. Using the same example I stated for the standard servo, applying a 1.75-ms PWM signal will cause a CR servo to rotate at approximately 15 rpm in the CW direction. You should note that the CR servo is not a very linear device due to the way the electronics control board functions, and you should not expect a precise relationship between the PWM pulse width and the output shaft speed. It will be close, but not exact. In any case, the CR servo's speed control will be more than adequate for this project.

It is now time to return to the platform build instructions now that you have an understanding of how CR servos work.

Robotic Car Power Supply

As you probably realize by now, the two CR servos driving the robot car require more voltage and current than can be provided by a Pyboard power supply. Fortunately, the Parallax Company provides a nice alternative power supply that will provide power to both the CR servos as well as to the Pyboard. In addition, its form factor matches quite nicely with the robotic car's chassis. Figure 7-5 shows this Li-ion battery power supply mounted on the robotic car chassis.

The Parallax part number for the power supply with two Li-ion batteries is 28989, and it has a built-in charging system for the two 18650-size Li-ion cells. Each cell provides 3.7 VDC at well over 1 A when fully charged. Two cells in series provide 7.4 VDC to the CR servos, which is plenty to allow them to be operated at maximum capacity if so desired.

I used four nonthreaded 1/2-inch nylon spacers with four 7/8-inch 4-40 machine screws to mount the battery supply to the car chassis. All four screws were secured with 4-40 threaded 1-inch nylon spacers, which hold the Pyboard mount plate. This mount plate holds the Pyboard and a small solderless breadboard. Figure 7-6 is a drawing of this Pyboard mount plate, including dimensions for all the holes required for the ½ inch 4-40 screws, which hold the plate to the nylon spacers.

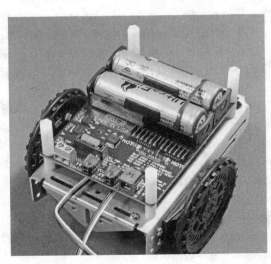

Figure 7-5 *Li-ion battery power supply mounted on the robot car chassis.*

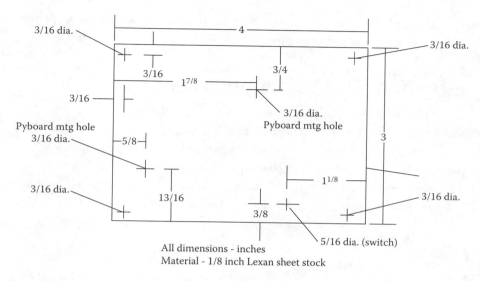

Figure 7-6 *Pyboard mount plate construction drawing.*

Figure 7-7 shows the Pyboard mount plate secured to the four nylon spacers along with the Pyboard itself and a small, solderless breadboard. I also plugged a Ping sensor into the breadboard. I will discuss this sensor in a later section. The Pyboard is attached to the mount plate using two 3/8-inch threaded, metal stand-offs. The solderless breadboard is attached to the mount plate using the double-sided tape that came with this board.

Figure 7-7 *Pyboard mount plate with the Pyboard, Ping sensor, and solderless breadboard attached.*

This completes the mechanical build instructions for the robotic car. I will next present the electrical and wiring instructions that will energize the car.

Electrical and Wiring Instructions

The electrical connections for the car are fairly simply and are shown in Figure 7-8, which is the entire electrical schematic for the robotic car.

The main power source for the two CR servos is the Li-ion power supply, which provides 7.4 VDC and has an energy capacity of approximately 2600 mAh when fully charged. This capacity should provide the two CR servos with well over six hours runtime before needing a recharge. Note that I included a toggle switch to

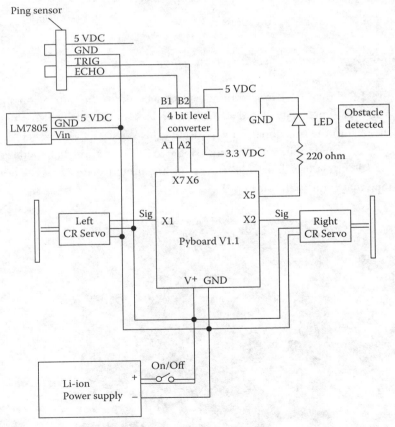

Figure 7-8 *Electrical schematic for the robot car.*

shut off the servo supply just in case you need an emergency stop. It is always wise to have an alternative method to securing motor power in a robot or similar device just in case something goes awry with the primary method of stopping the device, which I discuss in the software section.

The Pyboard is also powered by the Li-ion battery supply. It will draw approximately 100 mA, which is a minor load for the Li-ion power supply.

The Ping sensor that I mentioned earlier requires 5 VDC, which is provided by the Pyboard. The current draw for this sensor is about 20 mA, which is well within the 5 VDC external power supply specification for the Pyboard. The only other peripheral that draws any current is the light-emitting diode (LED) used to indicate when the robotic car has entered the obstacle avoidance routine. The LED draws less then 15 mA, which can easily be provided by the general-purpose input/output (GPIO) pin that controls the LED.

Ping Sensor

I purchased an HC-SR04 Ping sensor from Amazon.com. Figure 7-9 is a close-up view of this sensor.

The Ping sensor contains an embedded microprocessor as part of the sensor hardware. This processor controls the ultrasonic transmitter and receiver transducers that physically measure distance by bouncing discrete ultrasonic sound wave pulses off objects and timing how long the sound takes to travel from the sensor emitter to the sensor receiver. The distance is easily calculated because the speed of sound in air is relatively constant at 1,130 ft/sec. This method of

Figure 7-9 *HC-SR04 Ping sensor.*

operation is quite similar to how bats navigate in caves and attics. Figure 7-10 shows a block diagram for the sensor with all the principal components.

The sensor uses its own embedded processor to sense the distance between an obstacle and the robot car. This computed distance is then sent out as a series of digital pulses via a single wire to any processor that needs it, which in this case is the Pyboard. The processor in the Ping has a 10-µsec time measurement resolution, which translates to approximately a 1-inch distance measurement uncertainty in its total range of up to 100 inches. Of course, distance measurements also depend on the size and texture of the reflecting object. A wall provides excellent reflection, for example, whereas a stuffed animal would be more problematic.

Note that the Ping sensor requires 5 VDC for its power supply, and as a consequence the digital output pulse from the Ping sensor is also at a 5-VDC level. This level is incompatible with the Pyboard's maximum 3.3-VDC input level. This is the reason a level shifter chip is used as shown in the schematic in Figure 7-8. An inexpensive 4-bit level shifter mounted on a plug-in breakout board is available from adafruit.com as part number 757.

The digital communication protocol between the Ping sensor and Pyboard commences when the Pyboard (host device) generates a trigger pulse that causes the Ping sensor to emit a short burst of 40-kHz ultrasonic sound waves. This burst travels through the air, hits an object, and then bounces back to the Ping sensor. The Ping sensor will then provide an output pulse back to the host, which terminates when the echo is detected. Therefore, the width of this returned pulse is proportional to the distance to the target.

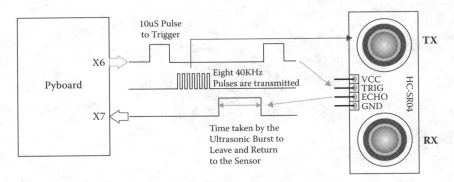

Figure 7-10 *Ping sensor block diagram.*

The electrical schematic in Figure 7-8 shows the interconnections between the Pyboard and the Ping sensor. Two wires connect the Ping sensor to the Pyboard through the level converter chip. One wire carries the pulse initiation trigger from the Pyboard, and the other wire sends the echo pulse back from the Ping sensor to the Pyboard. The duration between these two pulses is proportional to the distance between the sensor and the obstacle. The actual range measurement is converted within the Pyboard's MicroPython script, which I discuss in the software section on autonomous operation.

This section completes the physical robotic car assembly, including both the mechanical and electrical portions. It is now time to discuss the software that controls the car.

Robot Car Software

The first part of this section will be about the software that drives the CR servos. The second part will concern how the robot car may be operated in an autonomous mode to avoid objects while traversing a play field in a single direction.

CR Servo Control

The robotic car CR servos are controlled using the MicroPython PWM commands, which I first introduced in Chapter 2. If it has been awhile since you read that chapter, I would suggest you reread it, as it will help you understand the following discussion.

If you look at Figure 7-2, the upper servo is considered the left wheel and the bottom the right wheel. Referring to Figure 7-8, you will see that the right wheel servo is connected to pin X1. Similarly, the left wheel servo is connected to pin X2.

You will next need to set up a terminal session with the Pyboard as discussed in Chapter 1. Enter the following MicroPython commands, which will cause the servos to rotate at their maximum speeds, with the right wheel going in a CW direction and the left wheel in a CCW direction:

```
import pyb
s1 = pyb.Servo(1)
s2 = pyb.Servo(2)

s1.speed(100)
s2.speed(-100)
```

The CR servos rotate at approximately 30 rpm, which I discussed earlier in the chapter. The wheels are 2⅝ inches in diameter, which means the robot will travel forward approximately 8¼ inches per wheel revolution. At 30 rpm, the robot will traverse about 247.4 inches, or 20 feet 7 inches, in one minute. That is not terribly fast, only about 4 in/sec. I am fairly sure you will easily be able to keep up with your robot as you test it.

To stop the robot from moving, you must issue the following speed commands that place the servos in a neutral or nonrotating state:

```
s1.speed(5)
s2.speed(5)
```

These speed method arguments are set at 5 instead of 0 as you might expect to stop the servo rotating. This may due to a number of causes, but I suspect it is just the loose tolerances in the electronic components that make up the servo's electronic control boards.

By now, I think you probably figured out that the argument in these commands is simply a parameter that relates to the control pulse width. The value of 100 is equivalent to 2.0 ms and −100 is equivalent to 1.0 ms. Therefore, a total servo pulse width of 0.5 ms is represented by a 100 numerical range. However, all you really need to know is that the argument is only a percentage of the maximum speed. Hence, a value of ±50 represents the half-rotational maximum CR speed.

This last set of commands will cause the robotic car to go full speed in the reverse direction:

```
s1.speed(-100)
s2.speed(100)
```

Turning the robotic car is simply a matter of stopping the wheel, which is currently rotating in the direction you wish to turn. For instance, turning left would require you to issue these commands, assuming the robot was traveling forward at the time the command was issued:

```
import time
s1.speed(5)
time.sleep(1.6)
```

The servo is again restarted once the turn is completed to the desired degree of rotation. This situation raises an interesting control problem in that there is no feedback from the robot indicating how much rotation has actually occurred.

In purely control systems terminology, this is known as an open-loop control state. The only way to control the degree of rotation is by timing how long the nonrotating servo is stopped. Without getting into too much detail, it would take about 0.8 seconds to turn right 90 degrees assuming no slippage or skidding happens with the right wheel and that the robot was traveling at 4 in/sec forward at the time the turn was initiated. Of course, the actual rotation will depend on a variety of environmental conditions, including what type of surface the robot was traveling on, the presence of a slope or gradient, servo supply voltage, etc. It would probably be wise to slow the robot prior to entering a turn, in which case longer time durations could be invoked that would lead to more accurate turns.

Initial Test Run

The robotic car should now be ready for its first test run. The following MicroPython script will cause the robotic car to move forward for about 5 feet, turn around, and return to the starting point. First ensure that the on/off switch on the robotic car is set to off. Next, connect a universal serial bus (USB) cable to the car and enter this script into the main.py file, which is in the PYFLASH USB drive:

```
import time
import pyb

# Instantiate two servo objects
s1 = pyb.Servo(1)
s2 = pyb.Servo(2)

# Full speed forward
s1.speed(100)
s2.speed(-100)

// Travel about 5 ft.
time.sleep(8)

# Start a left turn
s1.speed(5)

# Left turn
time.sleep(1.6)

# Resume full speed back to the starting point
s1.speed(100)
```

```
# Travel back 5 ft.
time.sleep(8)

# All stop
s1.speed(5)
s2.speed(5)
```

You should next disconnect the USB cable and place the robotic car on a small wooden block so that the wheels do not touch the tabletop but still are free to spin. All you need to do is turn the car's power switch on, and you should observe the wheels immediately rotate at full speed for 8 seconds, and then the left wheel will stop for about 2 seconds, and finally both wheels will rotate at full speed for 8 seconds.

It is time to run the car on the floor once you are satisfied that the MicroPython script appears to run properly while on the block. Place the fully powered robotic car on the floor with at least 5 feet of clear space in front of the car. Simply turn on the power switch, and the robotic car should go forward in a straight line for about 5 feet, stop, and then turn around. It will then go back the same distance. The stopping point will be near the starting point; however, how close it comes to the starting point depends on the type of surface the robot is on and how much the robot actually turned. Feel free to experiment with the turn delay, as well as the straight-line delays, to see if you can actually stop the car at the starting point. I found a good deal of variability regarding the stopping point for the robot during my test runs. However, I was running the tests on a hardwood laminate floor and the robot's hard plastic wheels do not provide much friction, hence the large variability in the robot's finishing position.

The next portion of the robot software concerns the autonomous operational mode for the robot car.

Autonomous Mode

When you attempt to create obstacle avoidance software, you must think through all of the various scenarios that the robotic car may encounter in its travels from a starting to a finishing point. In order to make this a practical demonstration, I had to severely constrain the type of obstacle to use, as it would be impossible to account for all possible obstacle types. I chose a small cardboard box because it would be easy to place in the robot car's path and, if struck, should not cause any damage to the robot. The obstacle avoidance software must also be reasonably modifiable to accommodate other object types once the initial program has been demonstrated to be effective.

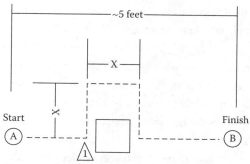

X = approximately 8 to 10 inches based upon
full power to the CR servos for 2 seconds.

Not to scale

Figure 7-11 *Playing field diagram.*

There really is not any formal procedure that will create an optimal obstacle avoidance algorithm. Consequently, I found that using a two-dimensional playing field with the starting and stopping points, as well as the obstacle placed in the robot path, helped me visualize how the robot could maneuver between the start and finish points while avoiding the obstacle. Figure 7-11 is a diagram of this sample playing field with the starting point marked as A and the finish point as B.

An obstacle detection point is shown as a numbered triangle. The robotic car path is shown as a dotted line with the X showing the predefined incremental path length. Table 7-1 is a pseudo-code listing that describes the robotic car's behavior as it starts on a straight-line path from A to B and encounters the obstacle placed near point number 1. Note that even with only one obstacle in the playing field, the code to implement the avoidance behavior rapidly becomes quite complex.

Playing Field Location	Obstacle No.	Behavior
Start point A	N/A	Straight-ahead motion on a direct path to finish point B for 8 secs
Obstacle detected	1	a. Turn left 90 degrees b. Move forward distance for 2 secs c. Turn right 90 degrees d. Move forward distance for 2 secs e. Turn right 90 degrees f. Move forward distance for 2 secs g. Turn left 90 degrees
On direct path	N/A	a. Move forward distance for 6 secs

Table 7-1 *Pseudo-Code for a Robot Car Obstacle Avoidance Algorithm*

All the steps to avoid an obstacle are contained in a conditional if statement. This program branch is taken when the ultrasonic sensor detects an obstacle within a threshold distance, which in this case is set at 25 cm, or 10 inches. This whole process of tracking the robotic car is called dead reckoning and presumes that all 90-degree turns are perfect and all traversed distances are in direct proportion to the preset time delays. In reality, this can never be the case, as the car tires will slip and motors do not instantly start and stop, meaning that the distances traveled will not quite match the desired values. The net effect of all these issues is that the car should arrive at a finish circle, where the circle diameter is proportional to the cumulative error in the car's path.

It is also convenient to have an indication of when the robot car enters the obstacle avoidance mode. This was easily accomplished by activating an LED connected to the Pyboard pin X5 whenever the distance measured is less than 10 inches. The following listing is the MicroPython code, which continuously measures the distance between the sensor and an obstacle and will raise a software flag when that distance falls below a 10-inch threshold.

```
from pyb import Pin

pinX5 = Pin('X5', Pin.OUT_PP)
pinX6 = Pin('X6', Pin.OUT_PP)
pinX7 = Pin('X7', Pin.IN)

micros = pyb.Timer(2, prescaler=83, period=0x3fffffff)

start = 0
end = 0

pinX5.low()

# This is a forever loop for debug purposes
# It is changed to a 0.1 sec loop in the actual code
while True:
    # Reset the counter to 0
    micros.counter(0)
    pyb.udelay(10)
    pinX6.high()
    pyb.udelay(10)
    pinX6.low()

    while not pinX7.value():
        # Get the starting echo count
        start = micros.counter()
```

```
while pinX7.value():
    # Get the finishing echo count
    end = micros.counter()

# This calculation is in cm's
if (end - start)/54 < 25:
    pinX5.high() # LED signal for obstacle

pyb.delay(20)
```

You should test this script with the ultrasonic sensor installed on the robotic car. Simply enter the code into main.py while the car is connected to a laptop. Then disconnect the laptop and switch on the car's own battery. Next monitor the LED as you place a reflecting object such as a book at various distances from the ultrasonic sensor. The LED should light if the book is placed closer than 10 inches, or about 25 cm, from the front of the sensor. If it appears to work okay, you will be ready to proceed to the next phase, which is incorporating the obstacle avoidance routine into this script.

The complete obstacle avoidance MicroPython script, which I named obstacle .py, is listed next. Notice that I placed both the obstacleCheck and obstacleAvoidance routines in separate method calls. This made it much easier to implement a simple loop that advances the robotic car in a straight path at approximately 0.5-second intervals while checking for obstacles in the path.

```
from pyb import Pin
import time
import pyb


# This is the obstacle checking routine
def obstacleCheck():

    micros = pyb.Timer(2, prescaler=83, period=0x3fffffff)

    start = 0
    end = 0

    pinX5.low()

    n = 50

    while n > 0:
        micros.counter(0)
        pyb.udelay(10)
        pinX6.high()
        pyb.udelay(10)
        pinX6.low()
```

```python
    while not pinX7.value():
        start = micros.counter()

    while pinX7.value():
        end = micros.counter()

    # Distance calculation is in cm's
    if (end - start)/54 < 25:
        pinX5.high()
        # Jump right to the obstacle avoidance routine
        obstacleAvoidance()

    pyb.delay(20)

    # This counter ensures the obstacle check runs for 0.1 sec
    n = n - 1


# This is the obstacle avoidance routine
def obstacleAvoidance():
    # Right turn
    s2.speed(5)
    time.sleep(0.8)

    # Fwd for 2 secs
    s2.speed(-100)
    time.sleep(2)

    # Left turn
    s1.speed(5)
    time.sleep(0.8)

    # Fwd for 2 secs
    s1.speed(100)
    time.sleep(2)

    # Left turn
    s1.speed(5)
    time.sleep(0.8)

    # Fwd for 2 secs
    s1.speed(100)
    time.sleep(2)

    # Right turn
    s2.speed(5)
    time.sleep(0.8)

    # Fwd
    s2.speed(-100)
```

```
    # Turn-off the obstacle LED
    pinX5.low()

    # Adjust global variable X for the obstacle avoidance distance
    X = X - 2

# Setup the pins used for obstacle detection
pinX5 = Pin('X5', Pin.OUT_PP)  # Flag signal
pinX6 = Pin('X6', Pin.OUT_PP)  # Trigger pulse to sensor
pinX7 = Pin('X7', Pin.IN)      # Echo pulse from sensor

# Setup the servo objects
s1 = pyb.Servo(1)
s2 = pyb.Servo(2)

# Start moving forward at full speed.
s1.speed(100)
s2.speed(-100)

# This loop just sends the car forward in 0.5 sec intervals
# Checks for obstacles every 0.5 secs

X = 0

while (X < 16):
    obstacleCheck()
    time.sleep(0.5)
    X = X + 1

# End of the direct path by time duration
# Stop the servos
s1.speed(5)
s2.speed(5)
```

In the next section I discuss how the obstacle avoidance algorithm actually worked, pointing out the successes and where some improvements could be made.

Obstacle Avoidance Demonstration

Figure 7-12 shows the robotic car set up on a real playing field with an obstacle in its direct path. The total distance between the start and stop points is 120 cm, or approximately 5 feet, so you need a moderate sized area to set up the course.

Turn on the robotic car to send it on its way. The car will simply proceed on a direct forward path until either an obstacle is detected or the path is traversed, which takes 8 seconds.

Figure 7-12   *Playing field with an obstacle.*

When the car entered the obstacle routine, I observed that the timed 90-degree turns were reasonably close, except some were a bit more due to the wheels skidding on the smooth laminate floor surface. I had mentioned previously that the playing surface on which I ran the robotic car can seriously affect its performance. I could have run the car on a carpeted surface, but that would slow the car down, which would then cause the turns to be much less than 90 degrees and subsequently cause the obstacle avoidance algorithm to fail. In addition, the car would not complete the desired path length because completing the total path depends on timed durations, not on actual distance traveled. These constraints and limitations can be overcome with additional experimentation and fine-tuning of the MicroPython script. However, the main intent of this project was to demonstrate how to build, program, and successfully operate a small robotic car, which I feel was completely met.

This last section completes this chapter. I do hope you will attempt to build this car, as it is great learning experience and a fun project.

Summary

This chapter focused on how to build and operate a robotic car, which was controlled by a Pyboard running MicroPython. In addition, a section on autonomous operation was included that dealt with a simple obstacle avoidance algorithm. Comprehensive build instructions were presented, as well as a section on how the CR analog servos worked. I also discussed the Ping ultrasonic sensor, which is used for obstacle detection. A simple demonstration of the robotic car on a playing field with an obstacle in its path concluded the chapter.

8

GPS with the Pyboard

Introduction

In this chapter, I will first discuss what constitutes the global positioning system (GPS) and the advantages and limitations that are involved with using the system. I will then explain how to connect a very capable GPS receiver to the Pyboard and subsequently process the received GPS sentences on the Pyboard.

Brief GPS History

The GPS was initially deployed in the early 1970s by the U.S. Department of Defense (DoD) to provide military users with precise location and time synchronization services. Civilian users could also access the system, but the services were purposefully degraded because the DoD was concerned with the risks that the system might help potential enemy activities. This purposeful degradation was lifted by order of President Reagan in the 1980s to allow civilians full and accurate GPS services.

The current GPS has 32 satellites in high orbits over the earth. Figure 8-1 shows a representative diagram of the satellite "constellation." The satellite orbits have been carefully designed to allow for a minimum of six satellites to be in the instantaneous field of view of a GPS user located anywhere on the surface of the earth. A minimum of four satellites must be viewed in order to obtain a location fix, as you will learn in the GPS basics section.

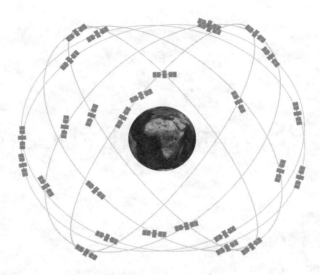

Figure 8-1 *GPS satellite constellation.*

Several other GPSs are also deployed:

GLONASS—the Russian GPS

Galileo—the European GPS

Compass—the Chinese GPS

IRNSS—the Indian Regional Navigation Satellite System

I will be using the U.S. GPS because vendors have made many inexpensive receivers available to purchase. All receivers function essentially the same way and conform to the National Marine Electronics Association (NMEA) standard discussed in a later section.

The Basics of How GPS Functions

I made up an analogous fictional position-location system to help explain how the GPS functions. First, imagine a 2 mile by 2 mile land area where this system is set up. The land terrain contains gently rolling "hills," with each no more than 30 feet in height. The subject using a "special" GPS receiver may be located anywhere within this area. Also located in this area are six 100-foot towers, each containing a beacon. The beacon atop each tower is configured to briefly flash a light and emit a loud sound burst simultaneously. Each beacon also emits the light and sound pulses once a minute, but as a specific time within the minute. Beacon one (B1) emits at the start of the minute, beacon two (B2) at 10 seconds past the start of the minute, beacon three (B3) at 10 seconds later, and so on for the remaining beacons.

It is also critical that the GPS receiver have a line of sight to each beacon and that the position of each beacon is recorded in an embedded database that is also constantly available to the receiver. The beacon positions B1 through B3 are recorded in terms of x and y coordinates in terms of miles from the origin that is in the upper-left corner as shown in Figure 8-2.

The actual position determination happens in the following fashion:

- At the start of the minute, B1 flashes and the receiver starts a timer that stops when the sound pulse is received. Because the light flash is essentially instantaneous, the time interval is proportional to the distance from the beacon. Sound travels at a nominally 1,100 feet/sec in air, so a 5-sec delay would represent 5,500-foot distance. The receiver must then be located somewhere on a 5,500-foot-radius sphere that is centered on beacon B1. Figure 8-3 illustrates this abstraction as a graphical representation taken from a Mathworks Matlab application.

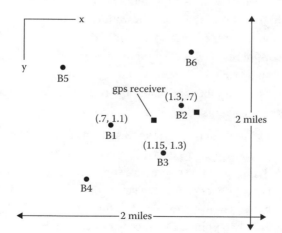

Figure 8-2 *Beacon test area.*

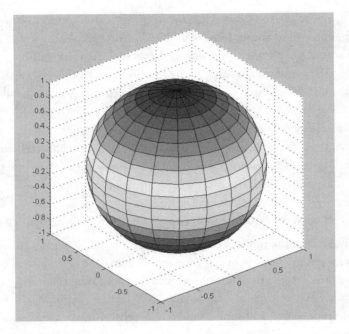

Figure 8-3   *One sphere.*

- B2 flashes next; suppose it takes 4 seconds for the B2 sound pulse to reach the GPS receiver. This delay represents a 4,400-foot sphere centered on B2. The B1 sphere and B2 sphere are shown intersecting in Figure 8-4. The heavily dashed line represents a portion of the circle that is the intersection of these two spheres. The receiver must lie somewhere on this circle. This circle is a straight line when observed in a planar or perpendicular view. There is still some doubt as to where the receiver is located on the circle. Thus, another beacon is still needed to resolve the uncertainty.

- B3 flashes next, and suppose it takes 3 seconds for the B3 sound pulse to reach the GPS receiver. This delay represents a 3,300-foot sphere centered on B3. The B1, B2, and B3 spheres are shown intersecting in Figure 8-5. The receiver must be located at the star shown in the figure. In reality, it could be either at a high or low point as the third sphere intersects the two other spheres at two points. The receiver's position has now been fixed with regard to x and y coordinates but not the third, or z coordinate. Guess what—you now need a fourth beacon to resolve whether the receiver is at the high or low point. I am not going to go through the whole process again, as I think you have figured it out by now.

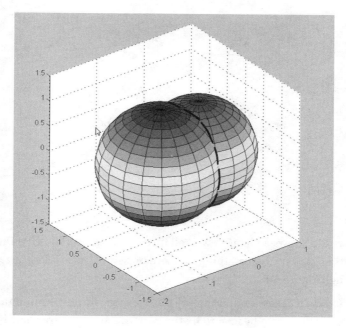

Figure 8-4   *Two spheres.*

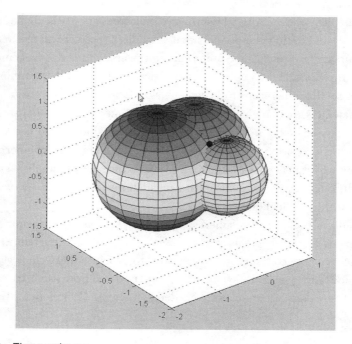

Figure 8-5   *Three spheres.*

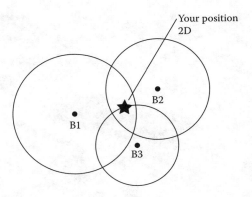

Figure 8-6 *Plane view.*

- Figure 8-6 shows a plane view of all three spheres with the GPS receiver position shown. You can think of it as a horizontal slice taken at z = 0 through Figure 8-5.

In summary, it takes a minimum of three beacons to determine the x and y coordinates and a fourth beacon to fix the z coordinate. Now translate beacons to satellites and x, y, and z coordinates to latitude, longitude, and altitude, and you have the basics of the real GPS.

The satellites transmit digital microwave radiofrequency (RF) signals that contain both ID and timing components that a real GPS receiver will use to calculate its position and altitude. The counterpart to the embedded database mentioned in my example is called an ephemeris, or celestial almanac, that contains all the data necessary for the receiver to calculate a particular satellite's orbital position. All GPS satellites are in high earth orbits, as mentioned in the history section, and are constantly changing position, which requires the receiver to use a dynamic means for their position fix, which in turn is provided by the ephemeris. This is one reason why it may take a while for a real GPS receiver to establish a lock, as it must go through a large amount of data calculations to determine actual satellite positions within its field of view.

The radii of the "location spheres" using my example are determined by the receiver using extremely precise timing signals contained in the satellite transmissions. Each satellite contains an atomic clock to generate these clock signals. All satellite clocks are constantly synchronized and updated from earth-based ground stations. These constant updates are needed to maintain GPS accuracy that would naturally degrade due to two relativistic effects. The best way to describe the first effect is to retell the paradox of the space-traveling twin.

Imagine there are two twins (male, female—it doesn't matter), one of which is slated to take a trip on a fast starship to our closest neighboring star, Alpha Centauri. This round trip will take about 10 years traveling nearly at the speed of light. The remaining twin will stay on earth awaiting the return of his or her sibling. The twin in the spaceship will accelerate very close to light speed and will patiently wait the 10 years it will take according to the clock in the ship to make the round trip. Now, according to Professor Einstein, if the traveling twin could view a clock on earth, he or she would observe time going by much quicker then was happening in the spaceship. This effect is part of the theory of special relativity and more specifically is called time dilation. The twin on earth would notice the clock in the spaceship turning much slower than the earth-bound clock. Imagine what happens when the traveling twin returns and finds that he or she is only 10 years older but the earth-bound twin is 50 years older due to time dilation. The space twin will have time-traveled a net 40 years into earth's future by taking the 10-year space trip!

The second effect is more complex than time dilation, and I will simply state what it is. According to Einstein's theory of general relativity, objects located close to massive objects, such as the earth, will have their clocks moving slower compared to objects that are farther away from the massive objects. This effect is due to the curvature of the space-time continuum predicted and experimentally verified by the general relativity theory.

Now back to the GPS satellites that are orbiting at 14,000 kph while the earth is rotating at a placid 1,666 kph. The relativistic time dilation due to the speed differences is approximately −7 μsec/day, whereas the difference due to space-time is +45 μsec/day, for a net satellite clock gain of 38 μsec/day. Although this error is nearly infinitesimal on a short-term basis, it would be very noticeable over a day. The total daily accumulated error would amount to a position error of 10 km, or 6.2 miles, essentially making GPS useless. That's why the earth ground stations constantly update and synchronize the GPS satellite atomic clocks.

NOTE *As a point of interest, the atomic clocks within the GPS satellites are deliberately slowed prior to launch to counteract the relativistic effects described earlier. Ground updates are still needed to ensure the clocks are synchronized to the desired 1-nanosecond accuracy.*

The Ultimate GPS Receiver

I will be using the Ultimate GPS receiver breakout board available for about $40 from Adafruit Industries. I first introduced you to this board in Chapter 3 where I was discussing and demonstrating the Universal Asynchronous Receive Transmit (UART) bit-serial interface. The GPS receiver is shown in Figure 8-7.

This receiver meets the following comprehensive technical specifications that make it ideal for this application:

- Satellites: 22 tracking, 66 searching
- Patch antenna size: 15 mm × 15 mm × 4 mm
- Update rate: 1 to 10 Hz
- Position accuracy: 1.8 meters
- Velocity accuracy: 0.1 meters/s
- Warm/cold start: 34 seconds
- Acquisition sensitivity: −145 dBm
- Tracking sensitivity: −165 dBm
- Maximum altitude for PA6H: tested at 27,000 meters
- Maximum velocity: 515 m/s

Figure 8-7 *Ultimate GPS receiver breakout board.*

- Vin range: 3.0 to 5.5 VDC
- MTK3339 operating current: 25 mA tracking, 20 mA current draw during navigation
- Output: NMEA 0183, 9600 baud default
- DGPS/WAAS/EGNOS supported
- FCC E911 compliance and AGPS support (offline mode: EPO valid up to 14 days)
- Up to 210 PRN channels
- Jammer detection and reduction
- Multipath detection and compensation
- Capable of attaching an external antenna
- UART for data communications (this feature will be further discussed in the following section on UART communications)

I will not need or use many of these features, but they are shown to give you an appreciation of the technical complexity and versatility for this particular GPS receiver.

Several key specifications are worth discussing a bit more. An acquisition sensitivity of −145 dBm means the receiver is extremely sensitive to picking up weak GPS signals. The −165-dBM tracking sensitivity means that the signal, once acquired, can lose up to 90 percent of its original strength yet remain locked in by the receiver.

Having an NMEA 0183 output operating at 9600 baud means the receiver generates standard GPS messages at a rate twice as fast as comparable receivers.

The Vin range of 3 to 5.5 VDC matches very nicely with the Pyboard 3.3-VDC operating voltage, thus eliminating the need for any buffer circuitry.

The 34-sec startup time is excellent and probably due in part to the extreme receiver sensitivity.

Initial GPS Receiver Test

It would be wise to check that the Ultimate GPS receiver is functioning as expected prior to running any code on the Pyboard. I have already discussed one initial test in Chapter 3 where I used a Raspberry Pi (RasPi) as the host microprocessor. If you do not have a RasPi, you can use the following procedure to test the GPS receiver with a laptop.

Figure 8-8 *External GPS antenna.*

First ensure that you have a good line of sight with the open sky to be able to receive the GPS satellite signals. I used an external GPS antenna because my test setup was indoors without any reliable satellite reception. The antenna was purchased from Adafruit Industries, part number 960, and is shown in Figure 8-8. It has a 5-m-long cable along with a magnetic base, making it easy to attach to any ferrous surface.

It is well worth the modest cost because erratic or unreliable satellite reception will quickly cause this project to fail. You will also need an antenna adapter to connect the external antenna's SMA connecter to the µFL connector situated on the Ultimate GPS receiver board. This adapter was also purchased from Adafruit, part number 851, and is shown in Figure 8-9. A word of caution: be very careful pushing the µFL connector on to the board because the inner pin appears to be quite fragile and probably would be damaged if excessive pressure is applied.

The quickest and easiest approach for a data connection is to temporarily connect the laptop to the GPS receiver with the USB/TTL cable using the connections

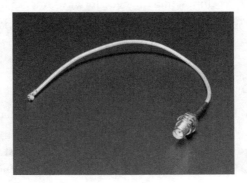

Figure 8-9 *SMA to µFL antenna adapter.*

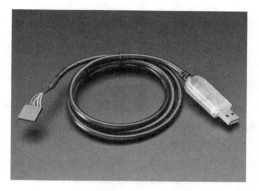

Figure 8-10 *USB/TTL cable connection from GPS to laptop.*

as shown in Figure 8-10. You can parallel-connect to the existing TXD and GND on the solderless breadboard without a problem.

I used a USB/TTL cable that I acquired from Adafruit as part number 70. It, too, is relatively inexpensive and can be reused in many other projects where serial USB communications are needed.

I next used the Terminal application installed on my Macbook Pro laptop with the baud rate set to 9600 to match the GPS receiver output. Figure 8-11 is a screen capture of the GPS data stream showing that the GPS receiver was properly functioning and receiving good satellite signals.

```
$GPGSA,A,3,30,05,13,20,21,,,,,,,,2.61,2.41,0.99*05
$GPRMC,013349.000,A,4314.3343,N,07102.7310,W,0.38,308.70,270716,,,A*76
$GPVTG,308.70,T,,M,0.38,N,0.71,K,A*3C
$GPGGA,013350.000,4314.3327,N,07102.7357,W,1,06,1.58,174.1,M,-32.8,M,,*52
$GPGSA,A,3,30,05,13,20,21,15,,,,,,,1.84,1.58,0.94*0F
$GPRMC,013350.000,A,4314.3327,N,07102.7357,W,0.94,291.00,270716,,,A*7F
$GPVTG,291.00,T,,M,0.94,N,1.74,K,A*38
$GPGGA,013351.000,4314.3323,N,07102.7375,W,1,06,1.58,173.0,M,-32.8,M,,*51
$GPGSA,A,3,30,05,13,20,21,15,,,,,,,1.84,1.58,0.94*0F
$GPRMC,013351.000,A,4314.3323,N,07102.7375,W,2.00,300.73,270716,,,A*78
$GPVTG,300.73,T,,M,2.00,N,3.70,K,A*3C
$GPGGA,013352.000,4314.3318,N,07102.7388,W,1,06,1.58,172.1,M,-32.8,M,,*58
$GPGSA,A,3,30,05,13,20,21,15,,,,,,,1.84,1.58,0.94*0F
$GPRMC,013352.000,A,4314.3318,N,07102.7388,W,1.71,296.28,270716,,,A*74
$GPVTG,296.28,T,,M,1.71,N,3.17,K,A*38
$GPGGA,013353.000,4314.3315,N,07102.7403,W,1,06,1.58,171.3,M,-32.8,M,,*51
$GPGSA,A,3,30,05,13,20,21,15,,,,,,,1.84,1.58,0.94*0F
$GPGSV,3,1,09,20,71,312,18,13,71,091,34,15,64,202,28,05,46,069,34*77
$GPGSV,3,2,09,21,39,307,14,02,17,135,,30,15,048,42,48,11,249,*74
$GPGSV,3,3,09,24,04,169,*4C
$GPRMC,013353.000,A,4314.3315,N,07102.7403,W,1.58,290.86,270716,,,A*75
$GPVTG,290.86,T,,M,1.58,N,2.92,K,A*3D
[]
```

Figure 8-11 *Terminal screen capture of GPS data stream.*

Once you have confirmed that the GPS receiver is operating properly, it is time to disconnect the USB/TTL cable and connect the GPS receiver to the Pyboard.

GPS Receiver UART Communication

The Ultimate GPS receiver uses a 9600-baud UART interface to communicate with the controlling microprocessor to both receive and transmit data back and forth. The UART interface pins on the Ultimate GPS receiver are shown in Figure 8-12. There are only three pins, including the ground or common, that are needed to establish communications between the GPS receiver and the Pyboard.

There is no need for a separate clock signal line because the UART protocol is designed to be "self-clocking."

Next you need to make the connections between the Pyboard pins and GPS receiver pins as shown in Figure 8-13. This figure is a duplicate of Figure 3-5 that I used in the Chapter 3 discussion on the UART protocol.

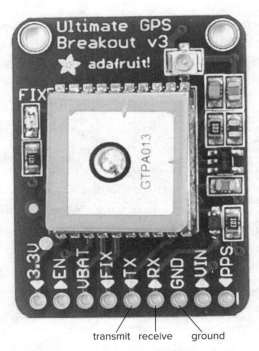

Figure 8-12 *Ultimate GPS data communication interface pins.*

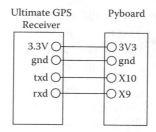

Figure 8-13 *Pyboard to GPS receiver connections.*

CAUTION   *Ensure TX from the GPS receiver connects to the Pyboard's X10 RX pin and likewise RX from the GPS receiver connects to the Pyboard's X9 TX pin. DO NOT CONNECT RX to RX or TX to TX even though that may seem the logical action to take. You will not damage anything, but data communications between the GPS receiver and the Pyboard cannot be established with those connections.*

Figure 8-14 shows the Pyboard connected to the Ultimate GPS receiver set up on an expanded solderless breadboard. Note that you will have to use an extra-wide solderless breadboard to accommodate the wide pin spacing of the Pyboard, if you chose to install regular header pins on the board as I did.

Figure 8-14 *Physical setup of the Ultimate GPS receiver and Pyboard.*

You will now need to create a read-eval-print-loop (REPL) session with your laptop in order to test the GPS receiver. The following REPL command sequence will both establish a UART object and start serial communications with the receiver and continuously display all the received GPS data strings.

```
MicroPython v1.7 on 2016-04-24; PYBv1.1 with STM32F405RG
Type "help()" for more information.
>>> from pyb import UART
>>> uart = UART(1, 9600)
>>> while True:
>>>      if uart,any():
>>>          print(chr(uart.readchar(), end='')
```

Note that those are two single quotes without a space between them after the end='' phrase in the print statement. Figure 8-15 is a screenshot of this continuous GPS data display. It scrolls by fairly quickly, making it quite difficult to read. You can stop the scrolling by pressing CTRL-C, which is the MicroPython keyboard interrupt key sequence.

It is also possible to capture single GPS sentences by using the procedure I first discussed in Chapter 3. The following REPL commands accomplish precisely that, and I have included an extensive discussion following the commands to help you understand what is happening.

```
MicroPython v1.7 on 2016-04-24; PYBv1.1 with STM32F405RG
Type "help()" for more information.
>>> from pyb import UART
>>> uart = UART(1, 9600)
>>> uart.init(9600, bits=8, stop=1, parity=None)
>>> myFrame = bytearray(63)
>>> uart.readinto(myFrame)
63
>>> myFrame
bytearray(b' $GPGGA, 032839.000, 4314.2943, N,
07102.6527, W, 1, 03, 5.37, 182.6, M, , *6A')
```

I now need to explain a few things regarding this command sequence. The command uart = UART(1, 9600) instantiates a UART object named uart, which uses UART port 1. The Pyboard pins associated with UART port 1 are X9 for receive (RX) and X10 for transmit (TX). The command also sets the baud rate at 9600 bits/second (bps).

The next command uart.init(9600, bits=8, stop=1, parity=None) initializes the UART protocol, fixing the number of data bits at 8 with 1 stop bit and no parity used. This serial configuration is a very common way most embedded asynchronous links are set up.

```
$GPGGA,215836.000,4314.3400,N,07102.6946,W,1,04,3.32,180.9,M,-32.8,M,,*5B
$GPGSA,A,3,02,06,23,09,,,,,,,,3.47,3.32,1.00*0D
$GPRMC,215836.000,A,4314.3400,N,07102.6946,W,0.21,329.25,270716,,,A*72
$GPVTG,329.25,T,,M,0.21,N,0.38,K,A*3A
$GPGGA,215837.000,4314.3398,N,07102.6953,W,1,04,3.32,180.8,M,-32.8,M,,*59
$GPGSA,A,3,02,06,23,09,,,,,,,,3.47,3.32,1.00*0D
$GPRMC,215837.000,A,4314.3398,N,07102.6953,W,0.18,278.54,270716,,,A*7B
$GPVTG,278.54,T,,M,0.18,N,0.33,K,A*38
$GPGGA,215838.000,4314.3397,N,07102.6961,W,1,04,3.32,180.8,M,-32.8,M,,*58
$GPGSA,A,3,02,06,23,09,,,,,,,,3.47,3.32,1.00*0D
$GPGSV,2,1,08,02,70,318,17,06,53,068,32,09,29,065,42,51,28,226,*7B
$GPGSV,2,2,08,25,26,300,,23,15,042,37,17,,,25,12,,,29*7F
$GPRMC,215838.000,A,4314.3397,N,07102.6961,W,0.37,266.76,270716,,,A*7B
$GPVTG,266.76,T,,M,0.37,N,0.69,K,A*35
$GPGGA,215839.000,4314.3398,N,07102.6968,W,1,04,3.32,180.7,M,-32.8,M,,*50
$GPGSA,A,3,02,06,23,09,,,,,,,,3.47,3.32,1.00*0D
$GPRMC,215839.000,A,4314.3398,N,07102.6968,W,0.50,274.55,270716,,,A*7F
$GPVTG,274.55,T,,M,0.50,N,0.93,K,A*33
$GPGGA,215840.000,4314.3398,N,07102.6973,W,1,04,3.32,180.6,M,-32.8,M,,*55
$GPGSA,A,3,02,06,23,09,,,,,,,,3.47,3.32,1.00*0D
$GPRMC,215840.000,A,4314.3398,N,07102.6973,W,0.40,277.18,270716,,,A*70
$GPVTG,277.18,T,,M,0.40,N,0.74,K,A*31
$GPGGA,215841.000,4314.3398,N,07102.6981,W,1,04,3.32,180.4,M,-32.8,M,,*5B
$GPGSA,A,3,02,06,23,09,,,,,,,,3.47,3.32,1.00*0D
$GPRMC,215841.000,A,4314.3398,N,07102.6981,W,0.60,267.01,270716,,,A*77
$GPVTG,267.01,T,,M,0.60,N,1.11,K,A*38
$GPGGA,215842.000,4314.3400,N,07102.6988,W,1,04,3.32,180.3,M,-32.8,M,,*50
$GPGSA,A,3,02,06,23,09,,,,,,,,3.47,3.32,1.00*0D
$GPRMC,215842.000,A,4314.3400,N,07102.6988,W,0.66,286.04,270716,,,A*77
$GPVTG,286.04,T,,M,0.66,N,1.22,K,A*34
$GPGGA,215843.000,4314.3400,N,07102.6991,W,1,04,3.32,180.2,M,-32.8,M,,*58
$GPGSA,A,3,02,06,23,09,,,,,,,,3.47,3.32,1.00*0D
$GPGSV,2,1,08,02,70,318,17,06,53,068,32,09,29,065,42,25,26,300,*73
$GPGSV,2,2,08,23,15,042,37,33,15,115,,17,,,24,12,,,30*77
$GPRMC,215843.000,A,4314.3400,N,07102.6991,W,0.54,285.17,270716,,,A*7E
$GPVTG,285.17,T,,M,0.54,N,1.00,K,A*34
$GPGGA,215844.000,4314.3399,N,07102.6992,W,1,04,3.32,180.0,M,-32.8,M,,*59
$GPGSA,A,3,02,06,23,09,,,,,,,,3.47,3.32,1.00*0D
$GPRMC,215844.000,A,4314.3399,N,07102.6992,W,0.37,265.52,270716,,,A*77
$GPVTG,265.52,T,,M,0.37,N,0.68,K,A*31
$GPGGA,215845.000,4314.3400,N,07102.6996,W,1,04,3.32,179.9,M,-32.8,M,,*54
$GPGSA,A,3,02,06,23,09,,,,,,,,3.47,3.32,1.00*0D
$GPRMC,215845.000,A,4314.3400,N,07102.6996,W,0.43,278.44,270716,,,A*7D
$GPVTG,278.44,T,,M,0.43,N,0.80,K,A*3F
$GPGGA,215846.000,4314.3401,N,07102.7001,W,1,04,3.32,179.8,M,-32.8,M,,*51
$GPGSA,A,3,02,06,23,09,,,,,,,,3.46,3.32,1.00*0C
$GPRMC,215846.000,A,4314.3401,N,07102.7001,W,0.53,284.74,270716,,,A*7B
$GPVTG,284.74,T,,M,0.53,N,0.99,K,A*36
$GPGGA,215847.000,4314.3399,N,07102.7014,W,1,04,3.32,179.7,M,-32.8,M,,*5D
$GPGSA,A,3,02,06,23,09,,,,,,,,3.47,3.32,1.00*0D
$GPRMC,215847.000,A,4314.3399,N,07102.7014,W,0.71,274.02,270716,,,A*75
$GPVTG,274.02,T,,M,0.71,N,1.32,K,A*38
$GPGGA,215848.000,4314.3398,N,07102.7027,W,1,04,3.32,179.5,M,-32.8,M,,*51
$GPGSA,A,3,02,06,23,09,,,,,,,,3.47,3.32,1.00*0D
$GPGSV,2,1,08,02,70,318,17,06,53,068,30,09,29,065,43,25,26,300,23*71
$GPGSV,2,2,08,23,15,042,37,33,15,115,,17,,,21,12,,,28*7B
$GPRMC,215848.000,A,4314.3398,N,07102.7027,W,0.83,276.85,270716,,,A*7B
$GPVTG,276.85,T,,M,0.83,N,1.53,K,A*3F
```

Figure 8-15 *GPS data screenshot.*

The next command `myFrame = bytearray(63)` creates a data "container" in which the raw GPS data will be stored. I found out through a bit of experimentation that 63 bytes seemed to be an optimal size for the bytearray object—at least for this initial test. I will actually be adjusting the size of the data container object later in the chapter when I discuss the parsing algorithm. You will note that

MicroPython returned the number 63, which indicates that it successfully allocated 63 bytes for this object.

The next command `uart.readinto(myFrame)` causes 63 bytes to be transferred from the Pyboard's UART receiver buffer into the `myFrame` bytearray. Note that this is an unsynchronized transfer, meaning that GPS sentences can be truncated or capture midstream. I had to repeat this command several times until I captured a complete GPS sentence, which is displayed by the next command.

The last command `myFrame` causes the content of the `myFrame` bytearray to be shown in the terminal window. I will not be discussing the actual GPS sentence content until later in the chapter.

You are almost ready to start using program scripts with the GPS receiver, but first I need to discuss the NMEA protocol and the messages that are being generated from the Ultimate GPS receiver.

NMEA Protocol

NMEA is an acronym for the National Marine Electronics Association, but nobody refers to it using the formal name. NMEA is the originator and continuing sponsor of the NMEA 0183 standard that defines, among other things, the electrical and physical standards to be used in GPS receivers. This standard specifies a series of messages that receivers use to create messages that conform to the following rules, also known as the Application Layer Protocol Rules:

- The starting character in each message is the dollar sign.

- The next five characters are composed of the talker ID (first two characters) and the message type (last three characters).

- All data fields that follow are delimited by commas.

- Unavailable data is designated by only the delimiting comma.

- The asterisk character immediately follows the last data field, but only if a checksum is applied.

- The checksum is a two-digit hexadecimal number that is calculated using a bitwise exclusive OR algorithm on all the data between the starting "$" character and the ending "*" character but including those characters.

A large variety of messages are available in the NMEA standard; however, the following subset is applicable to the GPS environment and is shown in Table 8-1. All GPS messages start with "GP."

Message Prefix	Meaning
AAM	Waypoint arrival alarm
ALM	Almanac data
APA	Auto pilot A sentence
APB	Auto pilot B sentence
BOD	Bearing origin to destination
BWC	Bearing using great circle route
DTM	Datum being used
GGA	Fix information
GLL	Lat/Lon data
GRS	GPS range residuals
GSA	Overall satellite data
GST	GPS pseudo-range noise statistics
GSV	Detailed satellite data
MSK	Send control for a beacon receiver
MSS	Beacon receiver status information
RMA	Recommended Loran data
RMB	Recommended navigation data for GPS
RMC	Recommended minimum data for GPS
RTE	Route message
TRF	Transit fix data
STN	Multiple data ID
VBW	Dual ground/water speed
VTG	Vector track a speed over the ground
WCV	Waypoint closure velocity (velocity made good)
WPL	Waypoint location information
XTC	Cross-track error
XTE	Measured cross-track error
ZTG	Zulu (UTC) time and time to go (to destination)
ZDA	Date and time

Table 8-1 *NMEA GPS Message Types*

Latitude and Longitude Formats

The two digits immediately to the left of the decimal point are whole minutes; to the right are decimals of minutes. The remaining digits to the left of the whole minutes are whole degrees.

Examples:

> 4224.50 is 42 degrees and 24.50 minutes, or 24 minutes, 30 seconds. ".50" of a minute is exactly 30 seconds.

> 7045.80 is 70 degrees and 45.80 minutes, or 45 minutes, 48 seconds. ".80" of a minute is exactly 48 seconds.

Parsed GPS Message

The following is an example of how to parse the received parsed GPGGA message that I previously showed you in the initial REPL session using a bytearray container:

```
$GPGGA, 032839.000, 4314.2943, N, 07102.6527, W, 1,  03,  5.37, 182.6, M, , *6A
1          2            3        4      5       6  7   8    9    10    11 12 13
```

1	GP	GPS NMEA designator
2	GGA	Fix message type
3	032829.000	UTC time
4	4314.2943	Current latitude: 43 degrees, 14 minutes, 18 seconds
5	N	North/South
6	07102.6527	Current longitude: 071 degrees, 02 minutes, 36 seconds
7	W	East/West
8	1	Fix quality:

0 = invalid

1 = GPS fix (SPS)

2 = DGPS fix

3 = PPS fix

4 = Real-time kinematic

5 = Float RTK

6 = Estimated (dead reckoning) (2.3 feature)

7 = Manual input mode

8 = Simulation mode

9	03	Number of satellites being tracked
10	5.37	Horizontal dilution of position (DOP)

DOP Value	Rating
<1	Ideal
1–2	Excellent
2–5	Good
5–10	Moderate
10–20	Fair
>20	Poor

11	182.6	Altitude
12	M	Meters
13	*6A	Checksum data

All GPS applications use some type of parser application to analyze data messages and extract the required information to meet system requirements. This will be discussed in the next section.

MicroPython GPS Parser

I will now be discussing a GPS parser named micropyGPS written in MicroPython, which was created by an open-source developer going by the user name "inmcm." The complete code is available at a Github website (https://github .com/inmcm/micropyGPS) from which the code can be downloaded or cloned. A comprehensive readme file is also included in the download, which provides excellent instructions on how to use the GPS parser with a Pyboard. Much of the following discussion is based on those instructions.

You must have the GPS receiver connected to the Pyboard before running the GPS parser application. The Pyboard must also be connected to the laptop with the PYFLASH directory opened. In addition, ensure that the GPS receiver is returning a good GPS data stream, or you will be frustrated in trying to use the parser. You should next follow these steps to start experimenting with the GPS parser:

1. Copy the micropyGPS.py file from the downloaded and expanded directory micropygps-master and put it into the USB PYFLASH directory.

2. Enter the following commands at the REPL prompt:
    ```
    >>> from micropyGPS import MicropyGPS
    >>> my_gps = MicropyGPS()
    >>> my_sentence = '$GPRMC,081836,A,3751.65,S,14507
    .36,E,000.0,360.0,130998,011.3,E*62'
    >>> for x in my_sentence:
    ... my_gps.update(x)
    ...
    'GPRMC'
    ```

NOTE   *You must physically enter the GPS sentence shown in the listing. It does not come from the UART interface.*

3. Next, try displaying some parsed data from the example GPS sentence:

```
>>> my_gps.latitude
(37, 51.65, 'S')
>>> my_gps.longitude
(145, 07.36, 'E')
>>> my_gps.timestamp
(8, 18, 36.0)
>>> my_gps.date
(22, 9, 05)
```

I would like to point out that the parser application is fairly constrained in that it only accepts GPS sentences that are put into a string variable, which in this case is named my_sentence. This means it is quite difficult to automatically stream data from the UART interface directly into the parser. I am sure that some clever open-source developer will eventually create code that accomplishes this task, enabling MicroPython to dynamically drive a liquid crystal display (LCD) or similar output device. However, for now you must either manually select the code to be parsed or use a bytearray container to initialize the GPS sentence to be parsed. I discuss both techniques in the next section.

Preparing GPS Sentences for Parsing

I will discuss two techniques to accomplish the process of initializing a GPS sentence for parsing. The first approach is entirely manual, but it also very efficient, and you can easily select the GPS sentence to be parsed. The second approach uses a script to capture a GPS sentence and then converts it into a form suitable for parsing. You do not have the option of preselecting the GPS sentence. You simply parse based upon what was received at the time you invoked the capture command. I call the two approaches manual and automatic parsing, respectively.

Manual Parsing

All you need to do for manual parsing is run the simple script that I first showed you earlier that continually scrolls the received GPS sentences. I have repeated the script here for your convenience:

```
>>> from pyb import UART
>>> uart = UART(1, 9600)
>>> while True:
>>>     if uart,any():
>>>         print(chr(uart.readchar(), end='')
```

Enter CTRL-C once you have displayed a good selection of GPS sentences. Next, copy the desired sentence into the copy/paste buffer using the appropriate method for your laptop's operating system. Ensure that you copy the entire sentence, including the opening and closing single quotes. Also verify that the checksum error value appears in the sentence. The parser application "checks" the checksum and will fail the parsing operation if it is either missing or incorrect. Now at the REPL prompt enter the following:

```
my_sentence = <pasted from the C/P buffer>
```

You need to next instantiate a parser object with the following commands:

```
>>> from micropyGPS import MicropyGPS
>>> my_gps = MicropyGPS()
```

All that's left is to parse the sentence using these two commands:

```
>>> for x in my_sentence:
... my_gps.update(x)
...
'GPGGA'
```

If all goes well, the parser will display the GPS message type, which in this case is 'GPGGA'. The update method called by the parser object examines each character in the GPS string and processes it according to the type of sentence encountered. At this point, the GPS parser will only parse the following sentences as detailed in Table 8-2.

The GPS sentence types currently handled by the parser application will naturally increase as open-source developers improve the application. Of course, nothing is guaranteed in the field of open-source development, but in my experience open-source developers take great pride in their work and typically will improve or enhance their applications.

Sentence Type	Description
GPRMC	Recommended minimum data for GPS
GPGLL	Lat/long data
GPGGA	Fix information
GPVTG	Vector track a speed over the ground
GPGSA	Overall satellite data
GPGSV	Detailed satellite data

Table 8-2 *Current GPS Sentence Types Parsed by MicropyGPS*

Automatic Parsing

I used the phrase "automatic parsing" to differentiate this approach from the much simpler cut-and-paste approach just presented. This approach uses a bytearray data container to first store the raw GPS sentence. I have shown these commands several times before, but I will display them once again for continuity. A UART object must first be created and then the current GPS sentence is loaded into the bytearray. These REPL commands accomplish this first portion:

```
>>> from pyb import UART
>>> uart = UART(1, 9600)
>>> uart.init(9600, bits=8, stop=1, parity=None)
>>> myFrame = bytearray(63)
>>> uart.readinto(myFrame)
63
```

MyFrame now contains a bytearray of the most recent GPS sentence received. You have no choice regarding the sentence type—it simply depends on what was being received at the time the command `uart.readinto(myFrame)` was executed. You should also realize that a bytearray data is not in a String format and you cannot directly load it into the GPS parser. Instead you must convert it to a String format and then select what is known as a substring to remove the bytearray wrapper information. The following commands accomplish those tasks:

```
my_sentence = str(myFrame)
my_sentence = my_sentence[13:75]
```

You can now use the converted and "substringed" my_sentence String with the parser in the following commands:

```
>>> for x in my_sentence:
... my_gps.update(x)
...
'GPGGA'
```

Of course, all of these commands may be incorporated into a single script, thus eliminating the manual cut-and-paste approach.

MicroPython GPS sentence parsing is still in what I consider a rudimentary phase; however, I believe I have presented sufficient details to allow you to pursue your own applications and perhaps develop far more sophisticated applications than I have presented.

Summary

I started the chapter with a brief history of the GPS, followed by a tutorial example that explained the basic underlying principles governing the system. The Ultimate GPS receiver was next discussed, focusing on the excellent receiver characteristics, as well as the easy serial communication link.

I discussed how to set up and test a serial console link using a USB/TTL cable as well as a serial terminal control programs for both a laptop and the Pyboard. The serial communication link between the GPS receiver and the Pyboard was set up and a communication tests was demonstrated to verify proper operation of all system components.

The NMEA 0183 protocol was thoroughly examined to illustrate the rich set of messages that are created by the GPS receiver. This project only uses a small subset of the data, but you should be aware what is potentially available. A parsed GPS message was also shown, along with a brief explanation of how to interpret lat/long data.

The remaining portion of the chapter concerned the MicropyGPS parser application, where I discussed two approaches to parsing real-time GPS data.

9

The ESP8266

In this chapter I will introduce you to the ESP8266, which is another inexpensive hardware platform that can run MicroPython. The ESP8266 also has a self-contained Wi-Fi circuit that permits network connections—a feature that is missing in the Pyboard. The ESP8266 is designed and manufactured by a Shanghai-based Chinese company, Espressif Systems. Most of the following module information comes from the company datasheet entitled "Expressif Smart Connectivity Platform: ESP8266." This datasheet is available in PDF format from the company website: https://espressif.com/.

I will not be presenting a single "big" project in this chapter, but will instead show you a series of experiments to demonstrate the ESP8266 capabilities. Much of the material in this chapter is based upon a great tutorial entitled "Quick Reference for the ESP8266." This tutorial is available from the website docs.micropython.org/en/v1.8/esp8266/esp8266/quickref.html.

The ESP8266 SMT ESP-12E Module

A core module for the ESP8266 named ESP-12E is available from adafruit.com as part number 2491B. Figure 9-1 is a top view of this module.

The actual size of the module is only about 1 × 5/8 in. You can also see a pcb trace Wi-Fi antenna on the left side of the module. An interesting feature of this antenna is that it is a quarter-wave dipole designed to resonant at 2.4 GHz.

Figure 9-1 *Top view of the ESP-12E module.*

One wavelength distance in meters for a 2.4-GHz frequency is calculated by the equation:

$$\lambda = v/f$$

where: λ = wavelength in meters

v = speed of light in meters/second

f = frequency in hertz

The wavelength calculation is:

$$\lambda = 300 * 10^6 / 2400 * 10^6 = 0.125 \text{ m}$$
$$\lambda/4 = 0.031 \text{ m} = 31 \text{ mm}$$

I measured the total length of the pcb trace and determined it to be 31 mm, not counting the length of the lead-in trace. This antenna length exactly matched the previous calculation. Figure 9-2 is a bottom view of the ESP-12E module.

Figure 9-2 *Bottom view of the ESP-12E module.*

All of the available module pins are clearly labeled in this figure. Table 9-1 provides details for all of these pins. The table rows are arranged to describe the pins situated on the left, right, and bottom sides, respectively.

There are a total of 16 GPIO pins, where 6 are multiplexed, meaning that multiple functions are provided by these pins depending on how they are configured. Ten dedicated GPIO pins should be more than sufficient for most projects; however, it is nice to know that six additional are available, provided that the additional functions are not required.

This concludes my brief introduction to the core module. The next section discusses the actual ESP8266 that will run the MicroPython language.

Pin Name	Function	Remarks
TXD0	Serial bit data	Output stream
RXD0	Serial bit data	Input data stream
GPIO5	General-purpose I/O	
GPIO4	General-purpose I/O	
GPIO0	General-purpose I/O	
GPIO2	General-purpose I/O	
GPIO15	General-purpose I/O	
GND	Ground	
RST	Reset	
ADC	Analog-to-digital conv	1-VDC maximum input voltage
EN	Enable	
GPIO16	General-purpose I/O	
GPIO14	General-purpose I/O	
GPIO12	General-purpose I/O	
GPIO13	General-purpose I/O	
VCC	Power supply	3.3 VDC nominal
SCLK	Serial clock	Part of the SPI communication protocol
MOSI	Master-out/slave-in	Part of the SPI communication protocol
GPIO10	General-purpose I/O	
MISO	Master-in/slave-out	Part of the SPI communication protocol
CS0	Chip select 0	Part of the SPI communication protocol

Table 9-1 *ESP8266-ESP12E Pin Descriptions*

The ESP8266 HUZZAH Breakout Board

The actual demonstration platform is an Adafruit ESP8266 HUZZAH breakout board. A top view of this board is shown in Figure 9-3.

Let me begin by stating I do not know why Adafruit chose this unusual name, other than the word "huzzah" is often connected to a greeting or exclamation of joy, similar to the interjections "hoorah" and "hooray." The availability of this inexpensive yet highly capable board is worthy of such a greeting. Incidentally, the Adafruit part number for the HUZZAH board is 2471, very similar to the ESP-12E module number, except it does not end with a B.

The figure shows the surface-mounted ESP-12E module soldered in place on the breakout board. The breakout board is so named because it extends all the core module pins to robust header pins, as well as providing additional support components to permit the core module to easily function on breadboards. A board schematic is available from the Adafruit website, which shows all the breakout board components. One key component is a low dropout (LDO) voltage regulator, which drops the 5-VDC power supply from the universal serial bus (USB) cable to 3.3 VDC required by the core module. In addition, there are two push button switches on the board. One is connected to the reset line, and the other to GPIO1. These two switches also serve a very important function in that they are used to switch the module into a programming mode such that MicroPython can be loaded. I will discuss that process in a later section.

A light-emitting diode (LED) is also onboard, which is permanently connected to GPIO1 and is very useful for running the classic "hello world" blink program.

Figure 9-3   *Top view of the HUZZAH board.*

Figure 9-4 *Bottom view of the HUZZAH board.*

Several protection diodes are also on the board to protect key pins, as well as several pull-up resistors for selected GPIO pins. Figure 9-4 shows a bottom view of the board.

As you can see from the figure, this board can accept an input power supply voltage level as high as 16 VDC. In addition, the RST and RX input voltage levels can be 5 VDC due to the protection diodes mentioned earlier. However, the logic level for most pins is a maximum of 3.3 VDC, just like the Pyboard. You must always be aware of this constraint and especially so when using many sensors, which are typically rated to provide 5-VDC digital output levels.

In the next section I will review some of the key hardware features of this board that you should be aware of prior to experimenting with it.

Key Hardware Technical ESP8266 Features

I will provide information in this section regarding important ESP8266 hardware technical specifications that are relevant to the types of projects that may be controlled with the MicroPython installed on the board.

CPU and Memory

The core CPU used in the ESP8266 is the Xtensa LX106, which is a 32-bit processor designed and manufactured by Tensilica, Inc. The Xtensa LX106 is a very interesting chip in that it has a "post-RISC" design. This means it does not utilize the classic ARM architecture, which is used in many other modern-day processors, including the core processor chip used in the Pyboard. Instead, it relies on a highly customizable core configuration where CPU components are added or subtracted from a framework to suit a particular application. The company also provides a software development toolkit that permits CPU designers to rapidly and accurately create a customized processor. This cuts the traditional CPU development time frame from years to just several months, which is a remarkable achievement. One very interesting outcome from this approach is that the internal word size for the LX106 processor is only 24 bits, whereas in a comparable ARM processor, it is 32 bits. This fact alone reduces chip complexity, improves instruction efficiency and saves power, which will be discussed in terms of power consumption later.

Firmware designed for an ARM processor is incompatible with the Xtensa series of processors. Any software created to run on an ARM chip must be recompiled from the source code to allow it run on an Xtensa CPU. Fortunately, there are several open-source firmware binaries available that will permit MicroPython to be loaded unto to the Xtensa LX106. I will discuss this process in a later section.

The module runs at an 80-MHz clock speed as purchased. It can be clocked up to 160 MHz using a simple MicroPython command. This means that a programmable phase lock loop (PLL) circuit must be incorporated into the clock circuit to permit this type of configuration. I have found that leaving the clock at 80 MHz is perfectly adequate for any operations that I did with the module. Slower clock speeds mean the chip runs cooler and uses much less power, which go a long way to improving reliability and chip life.

The internal Xtensa LX106 CPU memory is configured in what is known as a Harvard architecture where separate memories are allocated for instructions and data. The LX106 has 64 kB of instruction RAM and 96 kB of data RAM. These two memories constitute the dynamic memory where all programs are executed. There is also a 512-kB flash memory chip installed where the firmware and all user programs are stored. The combination of the CPU and memories are physically interconnected and stacked into what is known as a system-on-a-chip or SoC. The SoC is hidden under the metal can that forms a radiofrequency (RF) shield that is needed, in part, for Federal Communications Commission (FCC) certification.

Radio and Wi-Fi Hardware

The ESP8266 module contains a full-featured radio system that covers all the standard Wi-Fi channels between 2412 and 2484 MHz. It also features an approximate 20-dBm power output, which is remarkably strong and will easily encompass the typical Wi-Fi distance range of 100 m. The radio also handles the standard 802.11b, g, and n protocols without any difficulties. Bit rates in excess of 70 Mbps are easily accomplished when using the 802.11n protocol. All the applicable Wi-Fi modulation schemes are also handled automatically without the need for any specialized drivers or software. I will discuss how to use the radio subsystem in the network class discussion, which is presented a bit later in the chapter.

Support Circuits

A few support circuits are mounted on the module board, which are shown in the Figure 9-5 schematic.

A built-in LED is permanently connected to the GPIO0 pin. This makes it very easy to flash an LED if a quick and easy means is needed for a program debug indication.

Two push button switches are mounted on the board: one connected to the processor reset line and the other one that will ground the GPIO0 pin.

There is a low-voltage drop-out regulator (LDO), which will convert up to a 7-VDC input power supply to 3.3 VDC. The LDO can be powered by either a USB cable connecting the ESP8266 module to a laptop or a direct current (DC) supply connected to the VBAT pin. There are also two diodes in series with the power supply inputs to protect against reversed polarity being applied to the board.

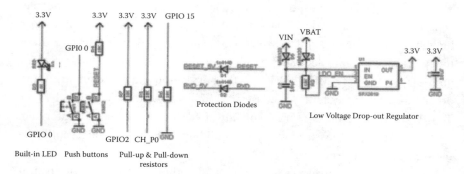

Figure 9-5 *Support circuit schematic.*

Finally, two overvoltage protection diodes are placed in series with the RXD and Reset line inputs to mitigate any problems with 5-V levels being applied on these lines.

This completes the brief hardware discussion. It is now time to examine the key software features of the ESP8266 module.

ESP8266 Software

I will start this section by stating that the firmware installed on the ESP8266 HUZZAH board will completely determine its nature and what you can do with it. The HUZZAH board as purchased from the Adafruit Company comes with NodeMCU (Lua for ESP8266) already installed on the board. This firmware will be replaced with the MicroPython firmware; however, those readers interested in the Lua language can go to https://nodemcu.readthedocs.io/en/master/ and read about this impressive language. The next section discusses where to obtain the latest ESP8266 MicroPython firmware and how to install it into the HUZZAH board.

Installing MicroPython on the HUZZAH ESP8266 Board

The latest version of the ESP8266 MicroPython firmware can be downloaded from the downloads section at micropython.org. At the time of this writing, the latest binary firmware that may be downloaded was esp8266-2016-08-01-v1.8.2-79-g0e4cae5.bin. This binary program is in turn loaded onto the HUZZAH board using a bootloader program named esptool.py.

The script (program) esptool.py is available at https://github.com/themadinventor/esptool. It is an open-source application designed specifically to load a bootable binary file into an ESP8266 board using the Python language. How esptool.py is used is dependent upon the host operating system (OS) that is running on the laptop. I used a MacBook Pro as the host laptop and consequently used the procedures for that system. If you use a Windows or Linux OS, I will refer you to Tony DiCola's fine Adafruit tutorial on how to flash the ESP8266. His tutorial may be found at https://learn.adafruit.com/building-and-running-micropython-on-the-esp8266/flash-firmware.

You will definitely need a current version of Python loaded onto your laptop in order to run the esptool.py utility. The following procedure is based on Tony's tutorial and will overwrite the current NodeMCU firmware and load MicroPython into the ESP8266 HUZZAH board.

You will first need to clone the source code. I opened a terminal window on the Mac and entered the following command to clone the source code:

```
git clone https://github.com/themadinventor/esptool.git
```

You now need to change into the new esptool directory that was created as a result of the clone operation:

```
cd esptool
```

Next run the script to check if it installed correctly. This operation will not start the flash operation, but simply checks that the script will actually run.

```
python esptool.py -h
```

You should see usage information displayed as follows:

```
usage: esptool [-h] [--port PORT] [--baud BAUD]
            {load_ram,dump_mem,read_mem,write_mem,write_
flash,run,image_info,make_image,elf2image,read_
mac,flash_id,read_flash,erase_flash}
                    . . .
. . .
```

In order to flash the firmware of an ESP8266 with the compiled MicroPython firmware, first make sure you have the esp8266-2016-08-01-v1.8.2-79-g0e4cae5.bin file copied into the esptool directory. Also ensure that the ESP8266 board is connected to the laptop with a USB cable. You will then need the serial port name for the device connection. In the case of the Mac, the connection was /dev/tty .usbserial-A1044AGY9. I determined the serial connection name by running the command `ls /dev | more` from the Mac terminal window.

The ESP8266 will next be required to be placed into its programming mode. You do this on the HUZZAH ESP8266 breakout board by pressing and holding the GPIO0 button and then pressing the reset button. Next release the reset button and then release the GPIO0 button. Note that Tony also discusses how to program the ESP8266 board in his tutorial.

You are now ready to run the following command, which uses the esptool.py script to upload the firmware:

```
python esptool.py -p SERIAL_PORT_NAME --baud 460800
write_flash --flash_size=8m 0 esp8266-2016-08-01-v1.8.2-
79-g0e4cae5.bin
```

where `SERIAL_PORT_NAME` is the name of the serial port the ESP8266 is connected to (like COM4, /dev/tty.serial-..., etc.) and `esp8266-2016-08-01-v1.8.2-79-g0e4cae5.bin` is the name of the binary bootloader file in the current directory. Note that you can have the binary file in a different directory, but you will need to provide the full path to where that file is located.

When esptool.py completes flashing the board, you should see the following displayed:

```
Connecting...
Erasing flash...
Writing at 0x0004d800... (100 %)

Leaving...
```

Once you see this text on the terminal window, you can be assured that the MicroPython firmware is now loaded into the ESP8266. Of course, you can always repeat this process if you wish to reload the firmware at any time in the future.

You are now ready to explore MicroPython with the ESP8266, which is discussed in the next section.

Exploring MicroPython Installed on the ESP8266 HUZZAH Board

First disconnect and reconnect the USB cable to start experimenting with Micro-Python with the ESP8266 board. Next open a terminal window or equivalent Windows application and connect to the board. The set of commands show how I started a read-eval-print-loop (REPL) session with the HUZZAH board:

```
screen /dev/tty.usbserial-A1044GY9 115200
```

Note that your serial port designation will differ from the one shown. Micro-Python should now respond as follows:

```
MicroPython v1.8.2-11-gbe313ea on 2016-07-13;
Type "help()" for more information.
>>>
```

The >>> prompt indicates that you are in a REPL session. You can now try out some simple interactive commands to prove that MicroPython is functioning properly:

```
>>> 8 * 6
48
>>> 9 / 2
4.5
```

```
>>> 9 % 2
1
>>> 8 / 3
2.666666
```

I will now show you a few methods from the machine class after successfully demonstrating the initial interactive REPL session.

Machine Class

The following interactive commands invoke several methods from the machine class, which is a higher-level abstraction modeling lower-level hardware features such as the processor clock speed:

```
>>> import machine
>>> machine.freq()
80000000
```

Note that no machine object was instantiated for this command sequence. This approach is often used when a software class is created as "static," meaning that only the class name is required in addition to the method name. The period between the class name and the method name is also referred to as the "member of" operator and is required to complete the entire command, as are the parentheses after the method name. In this case, the command machine.freq() returns the default clock value of 80,000,000, or 80 MHz.

The clock rate may also be reset by proving an argument in the freq() method. The next command resets the clock frequency to 160 MHz:

```
machine.freq(160000000)
```

Repeating the machine.freq() command should display the new frequency of 160 MHz. Incidentally, the hardware clock speed limit is also 160 MHz. Attempts to set a clock speed other than 80 or 160 MHz will result in the following error message:

```
>>> machine.freq(200000000)
Traceback (most recent call last):
  File "<stdin>", line 1, in <module>
ValueError: frequency can only be either 80Mhz or 160MHz
>>>
```

I will come back to the machine class later in the chapter when timers are discussed, but next I will introduce the esp class, which handles debugging messages, among other things.

Esp Class

Debugging messages are very useful for determining the status of the processor and various processes running on it.

```
>>> import esp
>>> esp.osdebug(0)
```

All debugging messages would then be directed through the UART 0 interface, which is the one normally used to connect with the host laptop. To disable the debug messages, simply issue this command:

```
>>> esp.osdebug(None)
```

I will next discuss the network class because that supports a principal ESP8266 feature of connecting to a Wi-Fi network.

Network Class

The following series of interactive commands will instantiate a network object, scan for available wireless networks, and then connect that object to the selected wireless network:

```
>>> import network
>>> wlan = network.WLAN(network.STA_IF)
>>> wlan.active(True)
>>> wlan.scan()
[(b'NETGEAR59', b'\x84\x1b^\x00\x88\x9f', 1, -67, 3, 0),
(b'linksys_SES_36356', b'\x00\x1d~\x1aMu', 6, -77, 3, 0),
(b'Mom_and_Dad_2', b'J\x94\xfc\x7f\xee\x84', 6, -88, 3, 0)]
>>> wlan.isconnected()
False
>>> wlan.connect('NETGEAR59', <password>)
>>> wlan.isconnected()
True
```

The harsh reality is that once a network connection is made, there are actually just a few limited actions that may be accomplished with such a connection. First and foremost, prevent yourself from running a browser application with this board. It simply is not designed to handle such an application because it has neither the memory nor processor capability for this. Instead, you should consider the network connection as a means for the wireless transfer of small amounts of data, probably through a simple socket connection. There is also an ongoing effort to implement the MQTT protocol on the ESP8266 MicroPython framework.

At the time of this writing, this effort is very much in development, but I am hopeful it will shortly come to a useful state. Please refer to my book, *The Internet of Things: Do-It-Yourself At-Home Projects for Arduino, Raspberry Pi, and Beagle-Bone Black* (McGraw-Hill Education, 2015), if you wish to learn more about MQTT and how it may be used with the Raspberry Pi.

If all goes well with the network setup, you will be ready to start testing some of the interesting peripherals that can be connected to the board.

Experiments

Figure 9-6 shows a solderless breadboard with an ESP8266 HUZZAH board mounted along with four of the experiments to be discussed. Do not be discouraged by the apparent complexity, as I will separately and carefully describe each experiment. I just wanted to show you that all four experiments could be mounted on one standard-size breadboard along with the ESP8266 module.

The first experiment will examine how to use the built-in analog-to-digital converter (ADC).

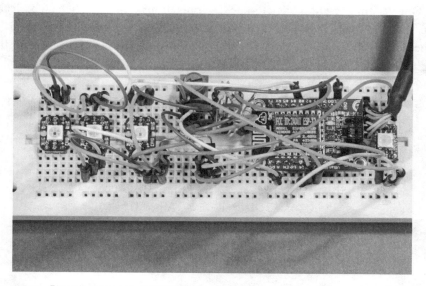

Figure 9-6 *Breadboard with ESP8266 and four experimental setups.*

Analog-to-Digital Converter

The ESP8266 contains a single 10-bit ADC that can accept up to a maximum 1-VDC input signal level. This means that you must be very careful to ensure you do not exceed this maximum level, or permanent damage will be done to the ADC. The first thing you must build is an input circuit that can provide a variable voltage between 0 and 1 VDC in order to test the ADC. Figure 9-7 shows the ADC test schematic where a printed circuit board (PCB) potentiometer can vary the input voltage.

The following command sequence will create an ADC object and set up the converter to measure input voltages:

```
>>> from machine import ADC
>>> adc = ADC(0)
>>> adc.read()
```

I next proceeded to apply a series of voltages to the ADC using the test circuit. The voltage levels were determined by reading a voltohmmeter (VOM) connected in parallel with the ADC input line. Table 9-2 shows the input voltage levels along with the corresponding digital ADC output values.

One very important parameter of ADC operations is linearity, which basically measures how closely the ADC output matches the changes in the analog input. One classic way to measure linearity is to plot the input voltage changes versus the digital output changes. Figure 9-8 is a plot of these changes.

Input Voltage	ADC Digital Output
0.0	5
0.1	105
0.2	213
0.3	312
0.4	418
0.5	522
0.6	622
0.7	726
0.8	834
0.9	934
0.988	1023

Table 9-2 *ADC Input Voltage and Output Digital Values*

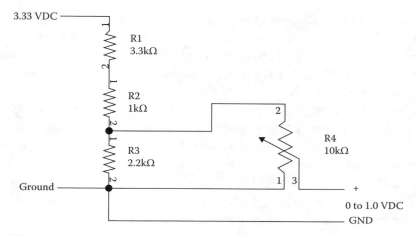

Figure 9-7 *ADC test circuit.*

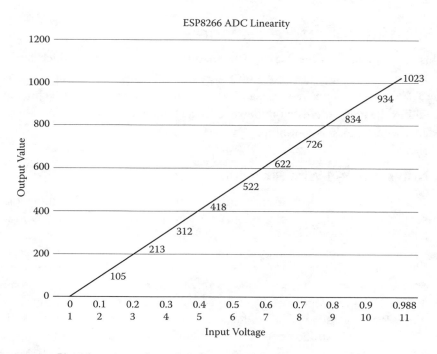

Figure 9-8 *Plot of analog voltage input versus ADC digital output.*

Linearity in this chart would be indicated by a constant slope of the plotted line. The line does indeed show a very constant slope, except for a slight downward curve at the extreme high end of the input voltage range. This would mean that there is a slight nonlinearity in the conversion process near the 1-VDC level, which is confirmed by the fact that the maximum analog voltage that could be applied is 0.988 VDC, resulting in the maximum 1023 digital output value. In a perfect linear situation, a 1.00-VDC input would cause the 1023 value.

Another key ADC specification is related to the conversion speed or, more specifically, samples per second (SPS). I did not directly test for this specification, but the ESP8266 manufacturer has stated in a Twitter entry that it is limited to 200 Hz or 200 SPS. Unfortunately, this is far too low to be of much benefit for any type of audio signal processing. However, this sample rate would be more than adequate for most sensor applications, such as ones used for measuring temperature and pressure.

The next peripheral I will discuss uses the 1-Wire data transmission protocol.

1-Wire Temperature Sensor

This experiment uses the Maxim DS18B20 temperature sensor, which implements the 1-Wire data transmission protocol. The following sidebar describes the 1-Wire protocol for those readers who are interested in how this technology functions. Readers interested in only using the temperature sensor can skip the sidebar without loss of continuity.

Notes on the 1-Wire Protocol

The 1-Wire protocol is the registered trademark name given to this digital serial communications protocol by the Dallas Semiconductor company when it first announced it in the early 1990s. However, any software implementing the protocol is not affected by the trademark. Dallas Semiconductor merged with Maxim years later but still manufactures 1-Wire products along with many other components. This is the reason why 1-Wire products have a DS prefix.

The 1-Wire system uses a master/slave, multidrop architecture with an open-drain connection, with pull-ups to using 3.3 to 5 VDC. This means that devices may be added or removed without any hardware configuration. All devices are discovered through software techniques. Also, all 1-Wire devices have a unique 64-bit identification number that is encoded in the ROM by the manufacturer.

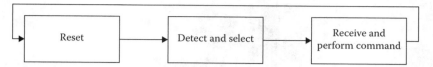

Figure 9-9 *1-Wire communication flow.*

The master and all slaves act as transceivers, meaning they can both transmit and receive but not at the same time. This mode of operation is known as half-duplex with data transmission being unidirectional. The master initiates all communication on the bus, with the slaves only responding to commands sent by the master. All data is sent by serial bits in a specific timed sequence. Bit timing is asynchronous with no external clock required because all timing is based on the signal transitions from the master.

All communication flow between the master and slaves involves three activities as shown in Figure 9-9.

In the first phase, the bus master must issue a reset command that synchronizes all elements on the 1-Wire bus. All slaves must respond to the reset, or the bus will not function as desired. A specific slave device is next selected to receive commands from the master. This selection is a multipart process that is started by discovering all the slaves currently connected on the bus using a binary search algorithm. Remember, all 1-Wire devices have a unique serial ID permanently programmed into their onboard ROM. The search algorithm reads all the IDs and records these values in a dynamic table hosted in the master. The master can then use a specific ID to send commands to a slave with that ID while all other slaves ignore the command.

The last phase involves having the master and selected slave engage in half-duplex communication where the master issues commands and the slave responds as designed.

Figure 9-10 shows the physical sensor, which is contained in a TO-92 transistor form factor. The sensor pin-out is also shown in the figure, which should greatly aid you in connecting it.

This sensor is actually a fairly complex device, as may be seen by reviewing the block diagram in Figure 9-11.

The following list details the key DS18B20 technical specifications:

- Configurable from 9 to 12 bits
- Measures −55° C to +125° C

Figure 9-10 *DS18B20 temperature sensor.*

- Measures −10° C to +85° C with a ±0.5° C accuracy
- Has its own 64-bit ID
- Powered from 3 to 5.5 VDC (may be parasitically powered)
- Converts 12 bits within 750 ms

This high level of device integration allows the sensor to be used with minimal software support, as will be demonstrated by the sensor commands. However, you will first need to connect the sensor to the ESP8266 using the Fritzing schematic shown in Figure 9-12.

The 1-Wire data line must also be pulled high by connecting a 4.7-kΩ resistor from the 3.3-VDC line to the data line.

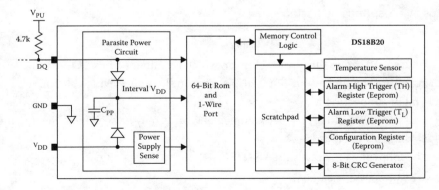

Figure 9-11 *DS18B20 block diagram.*

Figure 9-12 *DS18B20/ESP8266 interconnection schematic.*

The following REPL command sequence will instantiate a 1-Wire object, interrogate the sensor via the 1-Wire line, and finally display the sensed temperature in degrees Celsius:

```
>>> from machine import Pin
>>> import onewire
>>> ow = onewire.OneWire(Pin(14))
>>> import time
>>> ds = onewire.DS18B20(ow)
>>> roms = ds.scan()
>>> ds.convert_temp()
>>> time.sleep_ms(750)
>>> for rom in roms:
...     print(ds.read_temp(rom))
...
22.75
>>>
```

Note that the delay command is really only necessary if the command sequence is incorporated into a "forever" loop, which constantly communicates with the sensor. Of course, it is relatively easy to change the print statement to display the sensor in degrees Fahrenheit. The following statements do exactly that:

```
>>> for rom in roms:
...     print(9*ds.read_temp(rom)/5 + 32)
...
72.94998
```

This completes the 1-Wire sensor experiment. The next experiment is completely different and demonstrates the versatility of the ESP8266.

NeoPixel Demonstration

In this experiment, I will use three NeoPixel display devices to demonstrate how easy it is to control these LED display devices with the ESP8266. However, before showing you the experiment, I will first introduce you to the NeoPixel device.

The NeoPixel device I will use in this experiment is essentially a red-green-blue (RGB) LED with a built-in controller. Figure 9-13 is a close-up photograph of a single NeoPixel unit that is available from Adafruit as part number 1312. If you order that part number, you will receive a four pack with each NeoPixel mounted on a mini breadboard. All you need to do is solder two 3-pin headers to the PCB in order to use the NeoPixel device on a solderless breadboard.

The three LEDs contained in the NeoPixel are individually controlled by an internal chip named the WS2812. This chip is really a specialized microcontroller that has been programmed to accept serial bit data from a host microcontroller and light the individual LEDs according to the host's encoded data stream. The chip also receives and reshapes the serial bit stream so that it may be passed along to other NeoPixel devices downstream, thus forming a chainable link. The WS2812 and RGB LEDs are integrated into what is known as 5050 form factor. This form factor is a de facto standard for a variety of similar devices. Figure 9-14 shows the 5050 package outline along with a pin-out diagram.

Figure 9-13 *NeoPixel device.*

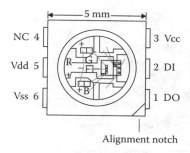

Figure 9-14   *5050 form factor and pin-out diagram.*

CAUTION   *The pin-out used in the NeoPixel PCB differs from the 5050 specification pin-out. I have detailed the pin-outs for both the 5050 specification and the NeoPixel PCB in Table 9-3. Also note that the NeoPixel device has Vcc tied to Vdd and to pin 4, whereas it is a No Connection on the 5050 module itself. You should also realize that the 5050 mounted on the NeoPixel is a standard unit—it just that the PCB traces are not routed in a one-to-one relationship.*

Figure 9-15 shows a Fritzing schematic that shows you how to wire up three 5050 RGB modules to an ESP8266.

The NeoPixels are wired up in a similar manner as the 5050 modules. I did not have a Fritzing symbol available for the NeoPixel device, which is the primary reason I used the 5050 symbol. You should not have any difficulty in wiring the breadboard as long as you connect all the grounds and Vcc's together and simply chain the DIs to DOs.

I also want to comment on why the LED power supply connection has been separated from the control power supply connection in the 5050 specification.

5050 Pin Number	NeoPixel Number	5050 Symbol	Description
1	1	DO	Control signal output
2	6	DI	Control signal input
3	3 and 4	Vcc	Control power supply
4		NC	No connection
5		Vdd	LED power supply
6	2 and 5	Vss	Ground

Table 9-3   *5050 and NeoPixel Pin-Out Details*

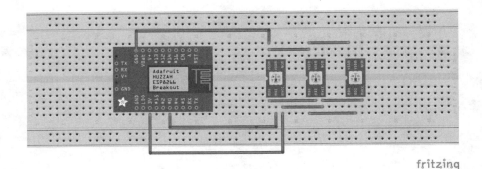

fritzing

Figure 9-15 *ESP8266/5050 RGB module interconnection schematic.*

It turns out that there are many different 5050 chained configurations, including some that are strips containing well over 100 LEDs. For example, the current demand of a strip containing 144 LEDs exceeds 8 A, which requires a beefy 5-VDC power supply.

You will be ready to enter the REPL commands to operate the NeoPixels once the hardware has been set up. Demonstrating the NeoPixels has been greatly simplified by the use of a cleverly crafted class named NeoPixel that hides all the intricate details of driving the actual devices. The following REPL sequence will turn on all the LEDs in the first NeoPixel device in the chain of three devices:

CAUTION   This command sequence will generate an intense white light from the first 5050 LED. Just be prepared for it.

```
>>> from machine import Pin
>>> from neopixel import NeoPixel
>>> pin = Pin(0,Pin.OUT)
>>> np = NeoPixel(pin, 8)
>>> np[0] = (255, 255, 255)
>>> np.write()
```

The command np[0] = (255, 255, 255) sets the intensity of the RGB LEDs in the first chained NeoPixel device. The range of values is from 0 to 255, which obviously means one data byte is used to set the intensity for each LED. The 0 index value in the np[0] array variable addresses the first device. This implies that up to 256 devices could be chained, which would be an impressive LED array.

The next series of commands will light the first device red, the second green, and the third one blue. In addition, each LED will be lit with a medium intensity.

You will also need the configuration commands such as the import and pin statements.

```
>>> np[0] = (64, 0, 0)
>>> np[1] = (0, 64, 0)
>>> np[2] = (0, 0, 64)
>>> np.write()
```

The next and final experiment demonstrates a similar LED display device, but it uses a different driver class.

APA102 Demonstration

This LED display device uses an SPI two-wire interface, instead of the much slower, single-wire NeoPixel interface. The LED device is also in a 5050 form factor. I used the Adafruit part number 2351, shown in Figure 9-16, for this experiment.

This LED display is called an APA102 5050 Cool White LED with Integrated Driver Chip, and it differs from the NeoPixel device in that it contains three white LEDs instead of red, green, and blue ones. The difference in display LEDs will make absolutely no difference in demonstrating how the device operates with the ESP8266.

What does make a big difference is that these SMT units only come unmounted. Adafruit also provides blank 5050 PCBs on which you can solder these devices. These blank PCBs are available in a ten pack with an Adafruit part number 1761.

Figure 9-16 *APA102 LED module.*

Figure 9-17 *Blank APA102 PCB board.*

Figure 9-17 shows one of these PCBs with a 5050 module held above the board with a tweezer.

It will be up to you to solder the APA102 5050 device to the blank board. You will definitely need a steady hand, as well as a needle-tipped soldering iron. I was finally able to solder one to a board after some difficulty and a few resoldering attempts. I would recommend that you first solder the two sets of 3-pin headers to the board and then plug the blank PCB into a solderless breadboard. Next, tape the 5050 device onto the board, carefully aligning the SMT solder pads on the device to the matching pads on the board. Then, very carefully solder a small "dab" to each pad. Figure 9-18 shows an assembled APA102 device.

Figure 9-18 *Assembled APA102 device.*

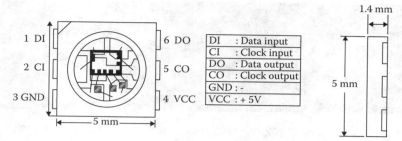

Figure 9-19 *APA102 pin-out and pin description.*

An APA102 device pin-out diagram and pin description is shown in Figure 9-19. Notice that it contains four control lines: two for clock in and out and two for data in and out. This configuration permits chaining APA102 devices in the same manner as was done for the NeoPixel devices.

You will be ready to enter the REPL commands to operate the APA102 device once the hardware has been set up. Demonstrating the APA102 has been greatly simplified by the use of another cleverly crafted class named APA102 that hides all the intricate details of driving the actual device. The following REPL sequence will turn on all the LEDs in the APA102 device:

CAUTION   *This command sequence will generate an intense white light from the 5050 LED. Just be prepared for it.*

```
>>> from machine import Pin
>>> from apa102 import APA102
>>> clock = Pin(14, Pin.OUT)
>>> data = Pin(13, Pin.OUT)
>>> apa = APA102(clock, data, 8)
>>> apa[0] = (255, 255, 255, 31)
>>> apa.write()
```

The next sequence of commands will dimly turn on each of the individual LEDs in the APA102 device. I photographed each LED activation and created a composite photograph just to demonstrate how precisely you can control this device. Figure 9-20 is the composite photograph that depicts the following command sequence:

```
>>> apa[0] = (2, 0, 0, 31)
>>> apa.write()
>>> apa[0] = (0, 2, 0, 31)
>>> apa.write()
>>> apa[0] = (0, 0, 2, 31)
>>> apa.write()
```

Figure 9-20 *Composite photograph of the individual APA102 LED activations.*

The bright spot in each of the photographs is an individual LED being activated. Normally you would not be able to distinguish the individual LEDs due to the effects of close proximity and combined brightness.

This last experiment completes my series of demonstrations on how to interface the ESP8266 module with various peripheral components. The next and final section deals with an experimental software application that allows you to create a REPL session using a network connection.

webREPL

This section deals with a beta application that permits you to conduct a REPL session using a web browser running on a host laptop. This app is entitled webREPL, which adequately describes its function of connecting to a MicroPython REPL prompt using a WebSocket connection. Perform the following steps to set up a webREPL session:

1. The first step is to download the host client from https://github.com/ micropython/webrepl.

2. Next establish a network connection to the ESP8266 access point as previously described in the earlier section on how to connect to a network.

3. Start the web client in the laptop and replace the default IP address with the ESP8266 IP address in the textbox shown upper-left corner. You should also retain the TCP port number originally displayed with the default IP address.

You should now have an active REPL session running over a web connection as shown in Figure 9-21.

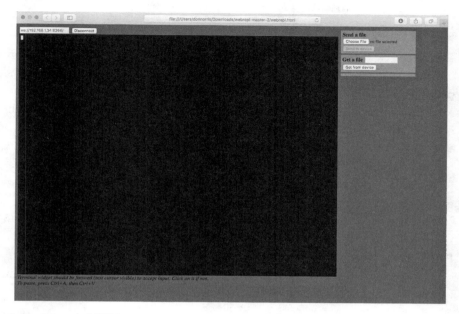

Figure 9-21 *webREPL session screenshot.*

The webREPL client also allows you to transfer files in both directions, which overcomes an important constraint when running MicroPython on the ESP8266. This feature will allow you to easily load large scripts into the module. In addition, you can automatically run a script upon boot-up by placing your script into the main.py, which is the normal MicroPython boot process. Figure 9-22 shows a file transfer in progress.

You should also realize that webREPL is very much a work in progress and there will be bugs and issues with the app. Hopefully, it will be improved over time to become a mature and steady app.

This last section concludes this chapter on the ESP8266 module.

Figure 9-22 *webREPL file transfer.*

Summary

I began the chapter with a brief introduction to the ESP8266 module, which is an inexpensive microcontroller incorporating a Wi-Fi radio system. Several variants of the ESP8266 modules are available, all of which may be programmed to run the MicroPython language. I chose to use the Adafruit ESP8266 HUZZAH breakout board as the chapter's demonstration board.

I next showed you how to load the MicroPython firmware into the HUZZAH board. A brief REPL session proved that the MicroPython program was successful. I did not go into an in-depth MicroPython discussion, as that was covered in earlier chapters.

The next sessions covered four experiments, which examined how simple it was to interface hardware peripherals to an ESP8266 system running the Micro-Python language.

The chapter concluded with a brief introduction to webREPL, which is an experimental application enabling you to initiate a remote REPL session that uses a browser client to connect with the ESP8266.

10

The WiPy

In this chapter I will introduce you to the WiPy, which is another relatively inexpensive hardware platform that can run MicroPython and connect to a wireless network in a similar manner to the ESP8266 discussed in the previous chapter. However, the WiPy is a much more capable platform than the ESP8266 because it has more memory, a fast processor, and more powerful network applications. The WiPy is designed and manufactured by Pycom, which is a U.K.-based company.

Please note that I will not be presenting a single "big" project or any experiments in this chapter, but will instead focus on the WiPy's networking capabilities, as that is the primary distinguishing characteristic that separates the WiPy from all other MicroPython-capable platforms.

Much of the following WiPy specification discussion comes from the Pycom website: www.pycom.io.

WiPy Specifications

Figure 10-1 is a top view of the WiPy module with most of the principal components noted.

Table 10-1 is a brief summary of most of the key WiPy specifications.

I will next present a brief introduction to the various Wi-Fi modes supported by the WiPy because they are important to understand in order to establish the appropriate wireless communication link with the module.

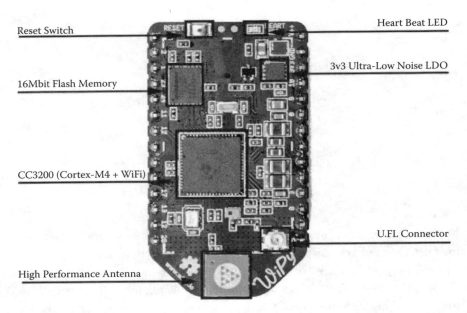

Reset Switch

Heart Beat LED

3v3 Ultra-Low Noise LDO

16Mbit Flash Memory

CC3200 (Cortex-M4 + WiFi)

U.FL Connector

High Performance Antenna

Figure 10-1 *WiPy top view with the key components.*

Feature	Quantity/Value	Remarks
Dimensions	1 × 1.77 inches	
Processor	1	Texas Instruments CC3200
Clock speed	80 MHz	
RAM	256 kB	
Flash	2 MB	
GPIO	25	3.3 VDC tolerant
UART	2	Async bit-serial interface
SPI	1	Sync bit-serial interface (HS)
I2C	1	Sync bit-serial interface (LS)
I2S	1	Audio data interface
SD/MMC	1	
Timers	4	16-bit
Wi-Fi	802.11b/g/n	16 Mbps
RTC	1	Real-time clock
SSL/TLS		Secure file transfer
WPA		Wi-Fi passkey

Table 10-1 *Key WiPy Specifications*

Wi-Fi Modes

The WiPy supports three basic Wi-Fi operational modes. I will discuss each one separately in the following sections.

Station

The station, or as it is more commonly referred to, the client mode, is the most commonly used Wi-Fi operational mode. In this mode, the module simply connects to an access point, which in turn provides all of the follow-on network connectivity. Common Wi-Fi routers typically provide access points for user local area networks (LANs) and are designed to provide any stations or clients with Internet Protocol (IP) addresses on demand. Of course, the user must provide at least the network ID [service set identifier (SSID)] and the appropriate passkey or password [wireless public key (WPK)] if the connecting network requires it.

A common form of the IP address that might be assigned could be 192.168.0.xx or 192.168.1.xx, where the precise number would depend on the router brand for the 0 or 1 in the third address segment and the xx segment is dependent upon the number of attached wireless devices to the particular router.

Access Point

In this mode, the WiPy is acting similar to the commercial routers mentioned in the previous section. The WiPy will initially provide an IP address to any device that attempts to connect to it as long as the appropriate SSID and WPK are provided. For the WiPy, these are

SSID: wipy-wlan
WPK: www.wipy.io

The IP address provided will be in the form 192.168.1.xx, where the xx in the fourth segment is dependent upon the number of attached wireless devices.

You should also not be confused with regard to logging into the WiPy filesystem. When you attempt that login, you will need to provide both a username and password as follows:

Username: `micro`
Password: `python`

You must also be aware that the WiPy access point does not incorporate a cable modem and cannot connect to the Internet through an Internet service provider (ISP). Therefore, the WiPy cannot provide any Internet connectivity to its stations or clients.

Direct

The direct Wi-Fi mode is also called the peer-to-peer connection. In this case, the WiPy will connect directly with another WiPy or similar device that supports a direct Wi-Fi connection. No router or other intermediate device is required or used for this connection type. The direct mode is similar to an older Wi-Fi connection called ad hoc. But unlike ad hoc, the direct mode implements a much easier and automatic approach to both discovery and network connections. I will not be demonstrating the direct mode in this chapter because the station and access point modes are all that I need to illustrate how the WiPy uses its networking capabilities. The next section discusses the WiPy expansion board, as that is required to be used prior to establishing an initial connection with the WiPy.

WiPy Expansion Board

Pycom also makes a WiPy expansion board, which is really the only practical way to use the WiPy. It is a reasonably priced accessory and is essential for using the WiPy. Figure 10-2 shows this board with a WiPy already mounted on it.

The expansion board has the following features:

- All of the WiPy pins are brought out to the header pins.
- There is a convenient micro USB socket mounted to supply the board with power and serial communications.
- An uncommitted user push button.
- Several light-emitting diodes (LEDs) for indicating board status and general use.
- Protected power supply with optional battery charging and operation.
- Micro SD card socket.
- Jumper header for enabling/disabling features.

These features and other useful devices and configurations are detailed in Figure 10-3.

Figure 10-2   *WiPy expansion board.*

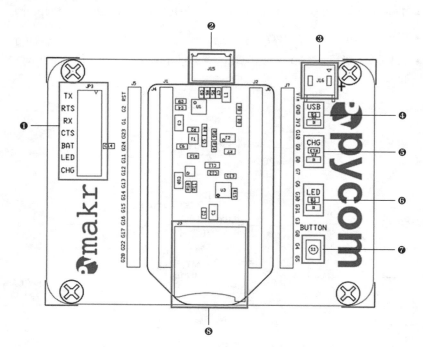

Figure 10-3   *Expansion board features outline.*

Feature Number	Description
1	Function selection jumpers
2	Micro USB socket for power and serial communications
3	LiPo battery connector (JST)
4	USB-powered LED
5	Charge indication LED
6	User LED
7	User push button
8	Micro SD card socket

Table 10-2 *Key Features on the WiPy Expansion Board*

Table 10-2 explains what each of these numbered features is and their purpose on the board.

The onboard power supply is also worthy of further explanation. Figure 10-4 is block diagram of the power supply and other board components.

The WiPy module may be powered from one of three sources, as you can see from the figure. These are

- The micro USB socket (whenever a cable is connected to a host laptop)

- The optional LiPo battery

- An external direct current (DC) power supply, 3.7 to 5.5 VDC, connected to Vin

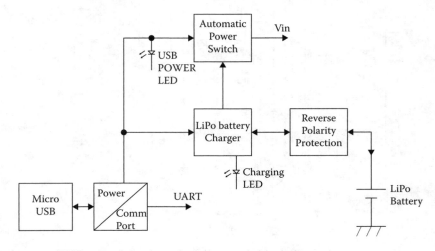

Figure 10-4 *WiPy expansion board power supply block diagram.*

Figure 10-5 *LiPo battery.*

There is also a protection circuit on the expansion board, which will help prevent problems such as attempting to connect a battery with reversed polarity. In addition, a circuit will automatically switch the power source between the battery and the micro USB socket.

Figure 10-5 is a picture of the optional LiPo battery, which is rated for 3.7 VDC and 350 mAh.

Note that the battery shown has a two-pin JST connector, which plugs into the matching JST socket mounted on the expansion board. A fully charged battery should power the WiPy for over 10 hours, but that time period depends a lot on how much wireless activity is happening during that timeframe. The Wi-Fi radio subsystem has a relatively large current usage compared to other WiPy subsystems.

The last expansion board feature I will discuss concerns the jumper header. Table 10-3 shows the jumper number and name along with which function is enabled or disabled, depending on if the respective jumper is or is not inserted.

The default state for all headers is to be removed, or out. You obviously need to insert a jumper at slot number 5 if you wish to use the optional battery feature. I would not recommend you use jumper slot 7, as a 450-mA charging current seems a bit high to me.

I mounted the WiPy and expansion board to a piece of clear 6 × 7–inch Lexan sheet plastic prior to running any of the following demonstrations or experiments.

Jumper No.	Jumper Name	Function (Jumper In)	Function (Jumper Out)
1	GPIO2	RxD	GPIO2
2	GPIO6	CTS	GPIO6
3	GPIO1	TxD	GPIO1
4	GPIO7	RTS	GPIO7
5	GPIO3	VBat	GPIO3
6	GPIO16	User LED enabled	User LED disabled
7	N/A	Battery charge at 450 mA	Battery charge at 100 mA

Table 10-3 *Expansion Board Jumper Header Configuration*

I also placed a full-sized solderless breadboard on the plastic sheet. Figure 10-6 shows this setup.

I have found it very useful to create this simple setup and also use manufactured jumper wires in the breadboard. In this way you can quickly configure and break down your experiments. I have also found I make fewer wiring errors using this technique, and it is simpler to troubleshoot, as all the circuit connection points are readily available. All in all, this is a highly recommended way to set up your experimental stations.

The next section concerns establishing an initial WiPy connection now that all the physical setups have been completed.

Figure 10-6 *WiPy experimental station.*

Creating the Initial WiPy Network Connection

I will start this section with the disclosure that I have based much of the following WiPy operational discussion on material that comes from the MicroPython website (https://micropython.org/resources/docs/en/latest/wipy/wipy/tutorial/intro .html). WiPy operation is distinct from either the Pyboard or the ESP8266 in that the initial connection to it must be done using a wireless mode, as the module does not come preconfigured to use a wired universal serial bus (USB) serial connection. I will show you how to create such a connection later in the chapter, but for now I will demonstrate the wireless method of connecting to a WiPy. I actually detailed how to accomplish most of this procedure in the brief section on the Wi-Fi access point mode of operation. Just follow these few steps to start your interactive session with the WiPy:

1. Connect a USB cable from your host laptop to the micro USB socket on the expansion board. Allow it a few seconds to fully boot.

2. Next scan the available Wi-Fi access points using your laptop. I can do this on my MacBook Pro host simply by clicking on the Wi-Fi icon in the upper-right corner. Laptops running Windows will differ.

3. Select the entry `wipy-wlan-0526`. Note the four numbers will differ for your WiPy.

4. Enter the WPA key www.wipy.io when the dialog popup appears. Once you enter this key, you will be connected to the WiPy in its access mode.

5. On the MacBook, I next started a terminal application session. This is needed to start a telnet communications session with the WiPy. On Windows hosts, you will need to initiate a PuTTY session. In either case, use 192.168.1.1 for the WiPy's IP address.

Figure 10-7 is a screenshot of my initial telnet session. In the figure you can see where I entered a username and password to obtain a WiPy read-eval-print-loop (REPL) prompt.

The password, which is `micro`, does not appear in the session text lines.

I entered a few arithmetic REPL commands just to demonstrate that the MicroPython language installed on the WiPy functions in a similar fashion to what has already been demonstrated on the Pyboard and on the ESP8266 HUZZAH board.

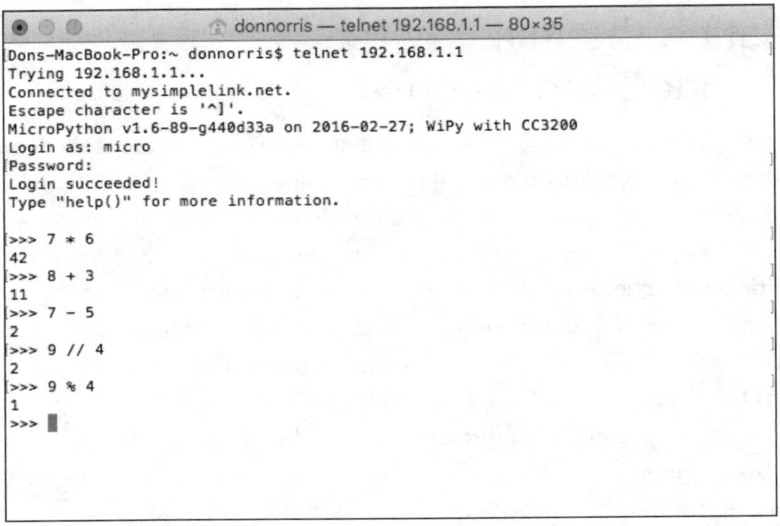

```
● ○ ○              ⚓ donnorris — telnet 192.168.1.1 — 80×35
Dons-MacBook-Pro:~ donnorris$ telnet 192.168.1.1
Trying 192.168.1.1...
Connected to mysimplelink.net.
Escape character is '^]'.
MicroPython v1.6-89-g440d33a on 2016-02-27; WiPy with CC3200
Login as: micro
Password:
Login succeeded!
Type "help()" for more information.

>>> 7 * 6
42
>>> 8 + 3
11
>>> 7 - 5
2
>>> 9 // 4
2
>>> 9 % 4
1
>>> ▊
```

Figure 10-7 *Initial telnet session.*

What is a bit different is my "odd" use of double forward slashes for the simple division of 9 by 4. Some readers may be familiar with this integer division notation. The two division operators simply instruct MicroPython to display the integer number of times 4 goes into 9. The answer is 2, but you are probably wondering why I invoked that peculiar division notation. The reason is that the MicroPython version running on the WiPy does not support floating point operations. Therefore, it is impossible to display the correct answer to 9/4, which, of course, is 2.5. The reason given in the WiPy documentation is that floating point support was left out due to memory space limitations. Attempting to perform a floating point operation will result in this error, `TypeError: unsupported type for operator`. There is also a further limitation in that the math library cannot be used because of the lack of floating point operations. My recommendation is that you avoid using the WiPy if your application requires any floating point calculations. Luckily, these types of calculations are often not needed in typical embedded, maker-type projects.

The telnet connection will not create a local USB drive on the host computer such as the one that appears when using the Pyboard. This was also an issue with the ESP8266. Lack of a "mirror"-type directory makes transferring files between the WiPy and the host more difficult than simply copying and pasting them, as would be the case for the Pyboard. Fortunately, the WiPy firmware has built-in support for the File Transfer Protocol (FTP). Setting up an FTP session is as simple as establishing a telnet session.

Follow these next steps to create an FTP session:

1. First ensure you are connected to the WiPy as an access point. Just repeat the steps previously discussed.

2. Enter the command ftp 192.168.1.1 to create an FTP session.

3. Enter the same username and password discussed in the telnet session instructions.

Figure 10-8 is a screenshot of an FTP session. In the figure you can see where I entered a username and password to obtain the FTP prompt.

Now, you will need to know how to use FTP commands to successfully run an FTP session. A lot of websites are available from which you can gather the information necessary to become proficient at using command-line FTP. To ease your burden a bit, I have summarized a few of the more useful FTP commands in Table 10-4. At least 42 FTP commands are available, but some are simply repeats of other FTP commands, just in a different form.

I created the following FTP demonstration to illustrate how simple and direct it is to use the FTP mode of operation to load your own script.

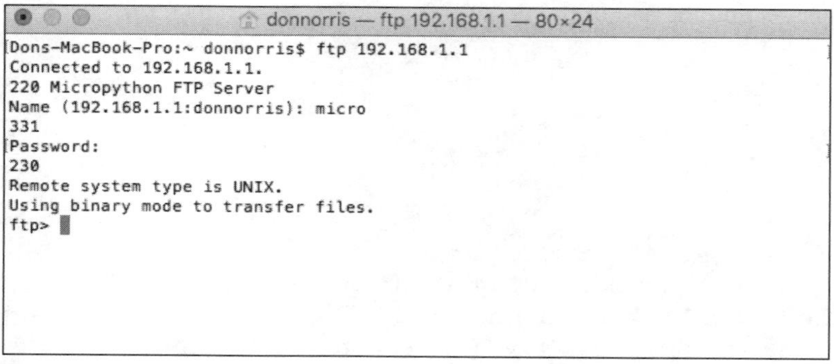

```
● ○ ○              🏠 donnorris — ftp 192.168.1.1 — 80×24
Dons-MacBook-Pro:~ donnorris$ ftp 192.168.1.1
Connected to 192.168.1.1.
220 Micropython FTP Server
Name (192.168.1.1:donnorris): micro
331
Password:
230
Remote system type is UNIX.
Using binary mode to transfer files.
ftp> █
```

Figure 10-8 *FTP session.*

Command	Description
bye	Exit the FTP application
cd	Change directory
delete	Delete a file
dir	List files in the current directory. Options: -C wide format -r reverse format -R list all files and subdirectories
get	Retrieve a file from remote computer
help	Display help screen
lcd	Display local directory
ls	Display list of local files
mkdir	Make a local directory
put	Send a file to remote computer
pwd	Display the working directory
rename	Rename a local file
rmdir	Remove a directory on remote computer
verbose	Toggle verbose mode

Table 10-4 *Useful FTP Commands*

FTP Demonstration

The following script was named blink.py and was uploaded to the WiPy using FTP. The script uses the existing heartbeat LED to blink, but changes its rate to once per second instead of the default: once every 4 seconds.

```
import time
from machine import Pin

led = Pin('GP25', mode=Pin.OUT)

while True:
    led(False) # turn off the heartbeat LED
    time.sleep_ms(1000) # wait one second
    led(True)  # turn on the heartbeat LED
    time.sleep_ms(1000) # wait one second
```

The following FTP command sequence transferred the blink.py file to the WiPy:

```
Dons-MacBook-Pro:~ donnorris$ ftp 192.168.1.1
Connected to 192.168.1.1.
220 Micropython FTP Server
Name (192.168.1.1:donnorris): micro
331
Password:
230
Remote system type is UNIX.
Using binary mode to transfer files.
ftp> cd flash
250
ftp> put blink.py
local: blink.py remote: blink.py
227 (192,168,1,1,7,232)
150
100% |********************************|   259       1.52 MiB/s    00:00 ETA
226
259 bytes sent in 00:00 (1.38 KiB/s)
ftp>
```

You should note that I had to change directories in the WiPy to put the blink.py in the flash directory where it can be accessed and run. This is exactly the same directory as used in the Pyboard. After you put the file in the directory, it is always a wise practice to list the directory contents just to confirm everything is set up as it should be. This next brief segment of commands shows that the blink.py file was successfully added to the flash directory:

```
ftp> cd flash
250
ftp> ls
227 (192,168,1,1,7,232)
150
-rw-rw-r--   1 root   root        34 Jan  1 00:00 main.py
drw-rw-r--   1 root   root         0 Jan  1 00:00 sys
drw-rw-r--   1 root   root         0 Jan  1 00:00 lib
drw-rw-r--   1 root   root         0 Jan  1 00:00 cert
-rw-rw-r--   1 root   root       167 Jan  1 00:02 boot.py
-rw-rw-r--   1 root   root       259 Jan  1 05:25 blink.py
226
ftp>
```

You will now have to exit the FTP application in order to run the blink.py file. You must start a new telnet session and enter the following commands to cause the heartbeat LED to blink at a faster rate:

```
Dons-MacBook-Pro:~ donnorris$ telnet 192.168.1.1
Trying 192.168.1.1...
Connected to mysimplelink.net.
Escape character is '^]'.
MicroPython v1.6-89-g440d33a on 2016-02-27; WiPy with
CC3200
Login as: micro
Password:
Login succeeded!
Type "help()" for more information.

>>> import blink.py
```

The last command import blink.py will cause the LED to start blinking at the fast rate. Just press CONTROL-C in the telnet window to stop the program.

You must place the script code into either the boot.py or main.py file if you want to have it run automatically upon boot-up. The WiPy boot sequence is identical to the Pyboard boot sequence—and, in fact, to all the MicroPython boot sequences. Any code discovered in the boot.py file is run first, followed by any code discovered in the main.py file.

It is time to show you a graphical user interface (GUI) application for file transfers now that I have shown you a basic command-line FTP application.

FileZilla

FileZilla is an open-source FTP application that uses a GUI for file transfers and other closely related tasks, such as directory and file maintenance. FileZilla runs on the host laptop, and the version appropriate to the host operating system (OS) may be downloaded from a variety of websites. This FTP application is very popular and is probably one of the most widely used. I elected not to initially introduce FileZilla, as I thought you should be aware of the important background tasks that are involved with the FTP process first. FileZilla hides these tasks and relegates the user to just pointing and clicking.

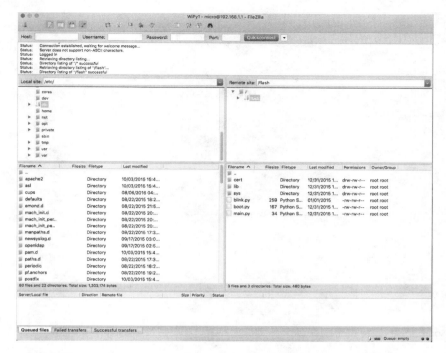

Figure 10-9   *Opening screenshot for a FileZilla FTP session with a WiPy.*

Figure 10-9 is a screenshot of the opening filescreen after I initially connect to the WiPy after opening the FileZilla application.

Please note that there is a Quick Connect button near the top menu bar. You should not use this, but instead open the Site Manager, which is found in the File drop-down menu selection. You will need to enter the information in this dialog box as shown in Figure 10-10.

Ensure that you enter 21 in the Port text box or you will not be able to establish a proper FTP connection.

File transfers using FileZilla are now simply a matter of dragging and dropping, as well as pointing and clicking. This is why it is such a popular application.

At this point, it is appropriate to switch the focus from the Wi-Fi access point operational mode to the station mode, where the WiPy becomes just another node in your LAN.

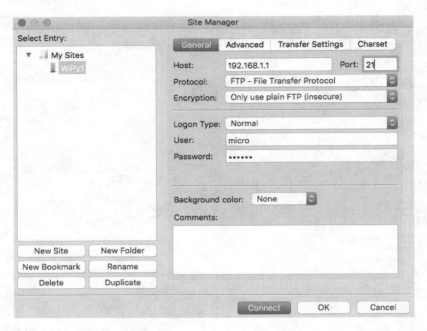

Figure 10-10   *Site Manager dialog box.*

Station Operation

The simplest approach to enable the WiPy to function in the Wi-Fi station mode is to first enable the USB serial connection, which I mentioned earlier in the chapter. The easiest way to do this is to put the following script named boot.py into the WiPy's flash directory by using FTP:

```
from machine import UART
import os
uart = UART(0, 115200)
os.dupterm(uart)
```

Then all you need to do is reboot the WiPy, and it will automatically run this script and instantiate a USB serial connection. Once that is created, just connect the WiPy using a micro USB cable to your laptop and run a terminal program appropriate to the host OS. I used a MacBook Pro and the built-in terminal program to establish a USB serial connection with the following command:

```
screen /dev/tty.usbserial-DQ005187 115200
```

Your device serial connection will differ from the one shown, as each host creates its own serial port identification. I saw the REPL prompt appear in the terminal window once I pressed the ENTER key. Now you are all set to put the WiPy into station mode and have it connected to your home network.

CAUTION   *Do not try to put the WiPy into the station mode when it is in the access point mode. It might appear to initially work, but it will eventually disconnect and fail to create a persistent station mode state. It is only applicable to a telnet session.*

You must first configure the WiPy into the station mode, as it always boots into the access point mode upon powering up. These next two REPL commands place the WiPy into the station mode:

```
>>> from network import WLAN
>>> wlan = WLAN(mode=WLAN.STA)
```

This next script will scan for all available wireless networks within the WiPy's range and automatically connect to the network preset in the script using the preset passkey:

```
>>> import machine
>>> nets = wlan.scan()
>>> for net in nets:
...     if net.ssid == 'yourwifissid':
...         print('Network found')
...         wlan.connect(net.ssid, auth=(net.
sec,'yourpasskey'), timeout=5000)
...         while not wlan.isconnected():
...             machine.idle()
...         print('WLAN connection successful')
...         break
...
Network found
```

This script can be either manually entered or placed in the boot.py file, where it will automatically be run when the WiPy is switched on.

Once the script is successfully run, the WiPy will simply become another node on your network with a dynamically assigned IP address, such as 192.168.1.34, which was the one assigned by my home router to the WiPy. In some cases, you may want to have a statically assigned IP address set to your WiPy. A static IP address will allow you to create network software, which will always "point" to the

WiPy, making it independent of any router assignment. The following script creates a static address for the WiPy. It is also always a good idea to assign a high-valued fourth-segment address to avoid any possible IP address conflicts with the IP addresses dynamically allocated by the router. In this example, I used 155 as the fourth segment. The router would have to allocate 154 IP addresses before a conflict happened, which is not likely to happen with any normal home network. This script would have to be in the boot.py file in order to have the operation function as expected.

```
import machine
from network import WLAN
wlan = WLAN() # instantiate a WLAN object without any presets

if machine.reset_cause() != machine.SOFT_RESET:
        wlan.init(WLAN.STA)
        # config statement must match the home network router settings
wlan.ifconfig(config=('192.168.1.155',255.255.255.0','192.168.1.1','8.8.8.8'))

if not wlan.isconnected():
        # change the placeholders to match your home network setting
wlan.connect('wifissid',auth=(WLAN.WPA2,'yourkey'),timeout=5000)
        while not wlan.isconnected():
        machine.idle() #conserve power while waiting
```

The two if statements in this script are important to prevent inadvertent disconnections while a telnet session is in progress.

This section concludes the WiPy network connectivity discussion. I will next proceed to discuss some important items regarding the boot process and the WiPy filesystem.

Boot Process and Restoring the Filesystem

I will start this section by stating it is just about impossible to write a program or execute a command that will permanently disable the WiPy. The WiPy design is inherently safe from inadvertent programming faults. You can always perform a safe boot in case you somehow cause the WiPy to enter a state from which either a soft boot (CONTROL-D) or a hard boot (press the reset button or power off then on) fails to correct the issue.

The safe boot procedure follows: When initially powering on the WiPy, any code found in the boot.py file will be executed first, followed by any script found in the main.py file. This normal boot sequence may be overridden by pulling

GP28 up (connecting it to the 3v3 output pin) during a reset. This procedure also allows you to revert to any one of three older firmware versions the WiPy holds, which include the factory firmware plus two user updates.

After a reset, and with GP28 still held high, the heartbeat LED will start flashing slowly. Releasing GP28 from the 3.3-VDC source within 3 seconds after the initial power-on will cause the WiPy to boot the most recent user firmware update. After 3 seconds with the GP28 pin still held high, the LED will start blinking a bit faster and the WiPy will select the previous user update to boot. If the previous user update is the desired firmware image, then GP28 must be released before 3 more seconds elapse. Three seconds later, if GP28 is still high, the factory firmware will be selected and the LED will flash quickly for 1.5 seconds. The WiPy will proceed to boot using the factory firmware.

The firmware selection mechanism is detailed in Table 10-5.

In all of these scenarios, safe boot mode is entered, which means that the execution of both boot.py and main.py is skipped. This safe boot is useful to recover from crash situations caused by the user scripts. The selection made during safe boot is not persistent; therefore, after the next normal reset, the latest firmware will run again.

Restoring the Filesystem

You may easily format the small WiPy filesystem in the unlikely case that it becomes corrupted. Formatting a filesystem will delete all files stored in internal WiPy memory, but not any files stored on the SD card. These two commands will format the internal WiPy memory:

```
import os
os.mkfs('/flash')
```

In addition to a fresh filesystem being created, both the boot.py and main.py files will be restored with their original factory contents.

Safe Boot Pin GP28 Released During:		
Time Frame	First 3-second window	Second 3-second window	Third and final 1.5-second window
Action	Safe boot, latest firmware is selected	Safe boot, previous user update is selected	Safe boot, previous user update is selected

Table 10-5 *Safe Boot Timing*

I will now change the topic a bit to focus on a nice development application provided by Pycom, which greatly helps in the building and testing of WiPy scripts.

Pymakr

Pymakr is a free integrated development environment (IDE) developed by Pycom, which will help you when creating scripts for your WiPy. IDEs have been around for a long time and are the favorite software development tools used by professional software developers. Having the Pymakr IDE available for the WiPy is quite fortunate and it is an opportunity that you should definitely take advantage of. Pymakr is somewhat akin to the Arduino IDE, which many readers are already well aware of and likely often use.

You will need to download and install the IDE from www.pycom.io prior to using it. Windows, Mac OS X, and Linux versions are available on the website. I found the download and installation process to be fairly quick and without any problems. Figure 10-11 is a screenshot of the application when it is first opened.

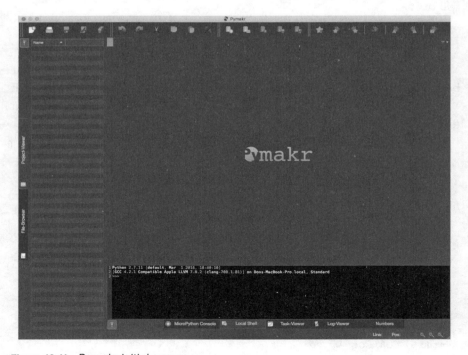

Figure 10-11 *Pymakr initial screen.*

You will first need to power the WiPy by using either a USB connected to your laptop or an external power supply, as discussed earlier in the chapter. Pymakr uses the wireless link to establish a telnet session with the WiPy. This means your laptop must be connected to the WiPy as an access point. Leaving your laptop connected to your regular network will result in the following Pymakr console error message:

```
Error while communicating with the MicroPython device:
Reattempting in 15 seconds...
```

This is an easy item to overlook and can be frustrating until you realize what is happening.

You will able to create projects in the Pymakr IDE, where you can both enter and edit code prior to running the code on the WiPy. You will still need to use an FTP application to transfer the new script to the WiPy because Pymakr does not incorporate FTP functions. I would advise you to think of Pymakr as a sophisticated code editor that has been tailored to only work with a WiPy module. There is nothing you can do with Pymakr that can't be accomplished using other external code editors. It is just that the Pymakr has been optimized to work with the WiPy. The choice on how to develop code is always a personal one.

Figure 10-12 shows the Pymakr screen with a script that I had downloaded from a Github site containing example scripts used with the Blynk mobile application.

Figure 10-12 *Pymakr screen with a script already loaded.*

This particular script is the actual Blynk library that is used for all Blynk projects. I will not be discussing Blynk any further, as that is a fairly complex application, but I did want you to be aware of its existence. Interested readers should look into Blynk, especially if they want to know how to control devices using a smartphone connected to a WiPy.

This last section on the Pymakr IDE concludes this WiPy chapter. There is much more to the WiPy than was possible to be covered in this brief discussion; however, I feel I have provided a reasonable introduction to this very interesting module that is continually evolving. Just go to the www.micropython.org and www.wipy.io sites to learn what is happening with the WiPy.

Summary

The chapter started with an introduction to the WiPy module, which is designed and manufactured by Pycom. A detailed list of key specifications was shown to illustrate the WiPy's capabilities.

I next discussed three Wi-Fi operational modes, as the WiPy uses two of the three for interactions with the outside world. In addition, the WiPy supports the UART interface, which allows you to establish a wired connection to the module using a USB serial link.

The WiPy expansion board was next covered, as that is an invaluable component to permit the WiPy to be practically used by experimenters. The board also permits three ways to power the WiPy, including the use of a LiPo battery, which I also reviewed.

I next demonstrated how to set up an initial telnet network connection to the WiPy using its access point mode of operation. This demonstration was immediately followed by an in-depth discussion of FTP, as that is the means by which files are transferred both to and from the WiPy.

A brief discussion on how to use the safe boot process followed. This process will restore the WiPy to a known "good" condition just in the remote chance that some errant code or script puts the WiPy in a poor or nonoperational state.

I ended the chapter with a brief discussion on the Pymakr, which is a free IDE available from Pycom that will help you in your code and script development.

11

Current and Future States of MicroPython

I will start this chapter on a somewhat depressing note in that what I discuss here will be outdated and obsolete by the time you read this in the published version. This situation is simply unavoidable where the technologies are so dynamic and rapidly changing. Significant changes occur with the MicroPython language and the associated hardware platforms such that time periods are measured in weeks, if not days. Nonetheless, I felt it was important to summarize at a certain point in time what the language status was and to describe the known hardware platforms and those immediately planned to be produced. Another bit of uncertainty is the fact that most, if not all, of the MicroPython efforts have been funded by a Kickstarter campaign. So far, all the direct and associated Kickstarter projects using this technology have been successfully funded—some with generous overcontributions, or stretch goals, as they are known in the Kickstarter world.

The MicroPython Language

As of fall 2016 the official MicroPython language version was at v1.8.2, with updates coming out daily as bugs are discovered and fixed as well as language enhancements added. The source code for the language is freely available on Github because it is open source and provided under the MIT license. You may also download and compile the source code to have your own most up-to-date version instead of waiting for the downloadable version to be made available on the www.micropython.org website. However, I would only recommend that course of action for readers who are well versed in this type of development.

Changes to the associated MicroPython libraries are constantly happening and are added to the Github site for inclusion in the overall download. One fortunate anchor in all these dynamic changes is that Python 3, on which MicroPython is based, is very stable with few changes that have to be reflected in the code base.

Documentation reflecting the current state of MicroPython seems to lag a bit from the base core already published and available from www.micropython.org. This is totally understandable, as the code developers are strictly volunteers and not full-time employees of a company whose profits are dependent, in part, on having up-to-date documentation. That is the reason why the micropython.org forums are so important. That is where you can ask questions about MicroPython and any of its associated hardware platforms and receive a prompt reply. However, remember the number-one rule regarding forum participation: you must do your own homework before asking the question. This means you must research all of the available documentation to see if an answer to your question exists before posting it. Nothing infuriates the forum experts more than someone who posts a question in the form of "How do I do this?" when the answer is clearly stated in the documentation. This type of question implies an unacceptable form of laziness on the poster's part and disrespect for the forum gurus, who the poster apparently considers some form of free help-desk support. I have stated this in the hope that my readers become intelligent and respectful forum participants. Who knows—after a while you, too, may become a forum expert.

There really isn't too much more to discuss regarding the upcoming changes to the MicroPython language. I really don't expect any disruptive changes to happen, but do expect continuous improvements both from optimization and expansion points of view (POVs). The optimization POV will hopefully be demonstrated in terms of more efficient language development that uses less memory and does more with its current allotment. The expansion POV concerns additional classes that allow easy interface implementations to both current and new sensors and devices that are not as yet supported.

Now it is time to list the current and anticipated hardware platforms that run MicroPython.

Hardware Platforms

Table 11-1 lists all the known hardware platforms that either currently run Micro-Python or are expected shortly to do so.

The only two boards shown in the table but not discussed in the earlier chapters are the LoPy and the SiPy. I will briefly discuss each of those boards, as they each bring some unique capabilities to the MicroPython community.

LoPy

The LoPy, just like the WiPy, was designed by pycom. It is essentially a WiPy with additional radio systems installed. Figure 11-1 shows the LoPy with two different LoRa antennas.

The radio systems incorporated into the LoPy according to advance pycom specifications are

* LoRa

* Wi-Fi

* Bluetooth

Technically, four radio systems are provided, if you count Bluetooth Classic and Bluetooth Low Energy (BLE) as two separate systems. However, pycom advertises the LoRa as a triple radio system module.

Name	Version	CPU	Memory (kB)		Main supplier
			RAM	Flash	
Pyboard	1.1	Cortex-M4	192	1024	micropython.org
ESP8266	2.0	Xtensa LX106	160	1024	Various suppliers: Adafruit AI-Thinker Qilianer Smarttime Sparkfun Wireless-Tag
WiPy	1.3	TI CC3200	256	2048	pycom
LoPy	–	Assumed TI CC3200	256	Up to 4096	pycom (available Q4 2016)
SiPy	–	Assumed TI CC3200	256	Up to 4096	pycom (available Q4 2016)

Table 11-1 *Current and Near-Future MicroPython Hardware Platforms*

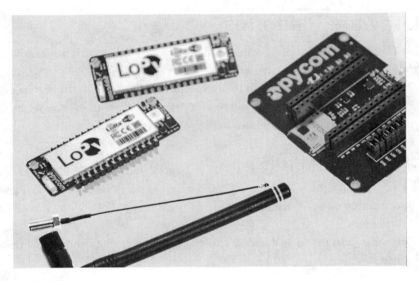

Figure 11-1   *LoPy.*

The LoPy gets its name from the inclusion of a LoRa radio system, which I assume is unfamiliar to most of my readers. That is the reason I provided a brief introduction to LoRa in the next section.

LoRa Radio System

LoRa is a digital radio subsystem designed to be part of a moderate-range wireless digital network named Low Power Radio Wide Area Network (LoRaWAN). LoRaWAN, in turn, is part of a larger specification known as Low Power Wide Area Network (LPWAN), which covers wireless battery-operated devices connected to regional, national, and even global networks. The focus of LoRaWAN is the Internet of Things (IOT), where it provides a standard to allow seamless communications between IOT devices covering a large area. Note that this approach is somewhat different than the approach taken by straight Wi-Fi connectivity where data from an IOT device is connected locally through a local area network (LAN) and then routed through the Internet to the appropriate destination. In a LoRaWAN situation, data is sent directly by digital radio (the LoRa) to a neighbor node, which might be its destination. If not, it is sent on, or "digipeated" to use an amateur radio term, to the next node until it reaches its final gateway destination. At that point the gateway created an Internet connection to reach the appropriate LoRaWAN server, which is how the network achieves its national and international reach.

LoRaWAN data rates are quite slow compared to Wi-Fi—only on the order of 0.3 to 50 kbps. However, this is adequate for most applications, as this type of network is designed to transfer sensor data and not audio and/or video information. However, low sample rate audio could easily be handled by the network if desired to supplement the raw digital data transfers. Figure 11-2 is a block diagram showing the principal components in a LoRaWAN communications link.

Notice the LoRa communications link notated in the lower-left corner of the figure. This is the LoRa radio link implemented in the LoPy. It is also easy to imagine that the LoPy could be loaded with LoRaWAN software to configure it as a sensor node as shown in the figure.

It is also quite possible to simply use two LoPys in a peer-to-peer arrangement, where data is directly transferred between the two devices. In that case, there would be no need for gateway or server nodes—just simple and direct communication between the two LoPys. The advantage of such a system is that it is possible to achieve up to a 21-km range if a line-of-sight (LOS) arrangement exists between the nodes. Otherwise, the LoRa specification states that 2 km is the maximum reliable range for a non-LOS condition.

I also would like to mention the frequencies that LoRa uses. They vary according to the region in which the system is used. Table 11-2 shows these frequencies.

These frequencies are in the Industrial, Scientific, and Medical (ISM) band and are unlicensed as long as the output power of the transmitters does not exceed a specified level.

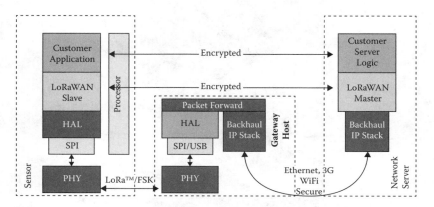

Figure 11-2 *LoRaWAN communications link.*

Region	Frequency (MHz)
Europe	868
North America	902
Australia, New Zealand, and South America	915 to 928

Table 11-2 *LoRa Frequencies*

SiPy

The SiPy is also designed by pycom and is almost identical to the LoPy. The SiPy is shown in Figure 11-3.

Figure 11-3 *SiPy.*

The SiPy implements the Sigfox radio protocol instead of the LoRa protocol, which is installed in the LoPy. Sigfox is a proprietary digital data transmission protocol developed by a French company named SIGFOX. This protocol has become popular in Europe in many IOT applications and is starting to become available in the United States at the start of 2016.

The next section compares Sigfox to LoRa in order to provide you with an understanding of these competitive technologies.

Sigfox versus LoRa

Sigfox is a narrow band, which uses a standard radio modulation called binary phase-shift keying (BPSK). BPSK takes very narrow spectrum portions and modulates the phase of the carrier radio wave according to the state of the binary data state (1 or 0). This modulation type enables the receiver to only listen in a tiny slice of overall spectrum, thus mitigating or minimizing noise effects.

LoRa uses a spread-spectrum modulation scheme that uses a wider amount of spectrum than Sigfox and consequently receives more noise or interference. However, the elevated noise due to a larger receiver bandwidth is mitigated because it's looking for a specific type of signal in the wide spectrum.

There are some significant cost differences between the devices used in Sigfox and LoRa networks. The LoRa nodes are generally less expensive because all the sensor nodes and gateway nodes can use the same technology for sending and receiving. The Sigfox receivers tend to be more expensive due to the technology involved.

There is one more item I wish to comment on. It may be anecdotal, but I did read that SIGFOX was claiming a 51-km LOS radio link could be established using their technology. That is quite remarkable if true, but I tend not to believe it to be achievable in a real-world situation.

Summary

I really didn't think this short chapter needed a summary, but I wanted to include one to essentially summarize the entire book from my point of view.

It is my sincere hope that after reading this book you have gained a reasonable appreciation of the MicroPython language and some of the associated hardware platforms that it runs on. I truly believe MicroPython will be a disruptive influence in the way embedded developers as well as makers, hobbyists, engineers, scientists, etc., will create new microcontroller projects. Shifting the focus from lower-level details to higher-level abstract concepts is the driving force behind MicroPython. It is all about making us more productive and effective in creating software applications that just work instead of grinding out all the miserable small details that always seem to arise in a project. Note that I am not saying that Micro-Python will eliminate all your debugging issues. What I am saying that as the language matures and more libraries are added to work with existing or new devices and sensors, MicroPython becomes an ideal tool to get these devices easily working in a project without having to sweat the small details.

INDEX